guide to cellular e

A series of books in biology

lynne c. carter
university of california berkeley

guide to cellular energetics

W. H. Freeman and Company san francisco

Printed in the United States of America

International Standard Book Number: 0-7167-0598-2

Library of Congress Catalog Card Number: 72-5397

contents

acknowledgment

The author wishes to express sincere thanks to the following persons for their assistance in the development of this book: Roderic B. Park, Frederick Reif, John E. Hearst, James B. Ifft, Helene Fatt, Diane Robertson, Marvalee H. Wake, and M. I. Charles E. Woodson. The author was partially supported by a fellowship from the National Science Foundation during part of the preparation of the manuscript.

about this book

This book is a supplement to the lectures on cellular energetics and cellular metabolism usually included in introductory biology, cell physiology, and molecular biology courses and related undergraduate courses. It provides an opportunity to learn or to review some basic chemistry concepts that are needed to understand cellular metabolism.

The material is arranged in the format of programmed learning, the most noticeable feature of which is that you are expected to answer questions interspersed throughout the subject matter. Although this procedure takes slightly more time than simply reading a textbook, you will find that you are likely to learn the material more readily if you respond to it as you go along. In this book you will be able to skip over material you already know and to proceed more slowly through subjects that are new to you, so that you will get a learning prescription based on your individual needs.

The directions for working through the book will be given later. First, we will take an overall look at cellular metabolism.

an overview of cellular metabolism

Organisms require food, whether they obtain it directly from their environment or build their own food molecules from simple substances as the green plants do. In the cell, food molecules may be utilized in either of two ways. A food molecule, or part of a food molecule, may be used as a building block in the manufacture of new cell parts or the repair of old ones. The second possible fate, which will be the main concern of this book, is that the food molecule may undergo a series of reactions whereby it is broken down into simpler substances.* In some cells these simpler substances undergo a second series of reactions during which they react with oxygen and are converted to the even simpler products, carbon dioxide and water. (This process is called respiration.) The simple molecules or the carbon dioxide and water that are produced by the breakdown of food molecules are usually excreted by the cell.

What benefit does the cell derive from the breakdown of food molecules? To find the answer to this question we have to consider

*This breakdown of food molecules in the cell should not be confused with the process of digestion, which occurs earlier, usually in the digestive tract.

that almost all cellular activities depend on chemical reactions. Here are three examples of such reactions for you to look at briefly.

1. *Growth.* Growth involves construction of new cell material. Macromolecules must be synthesized from smaller subunits called monomers by the formation of chemical bonds joining the monomers. For example:

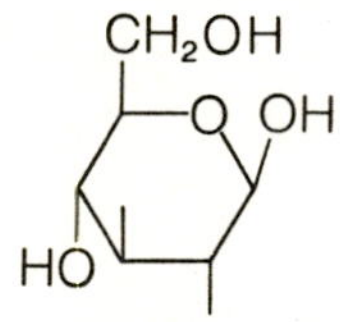

Monomer (glucose)

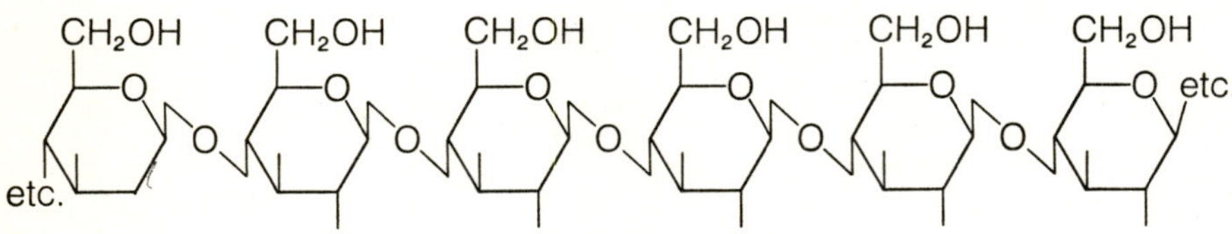

Macromolecule (cellulose—contains in the range of 2,000 glucose monomers)

2. *Regulation of the concentration of ions inside the cell.* It has been proposed that carrier molecules are the mechanism by which the cell regulates ion concentration. According to this model, chemical bonds must be formed between the carrier molecules and ions; later these bonds must be broken. The six diagrams that follow show how an ion might be excreted from a cell by a carrier molecule.

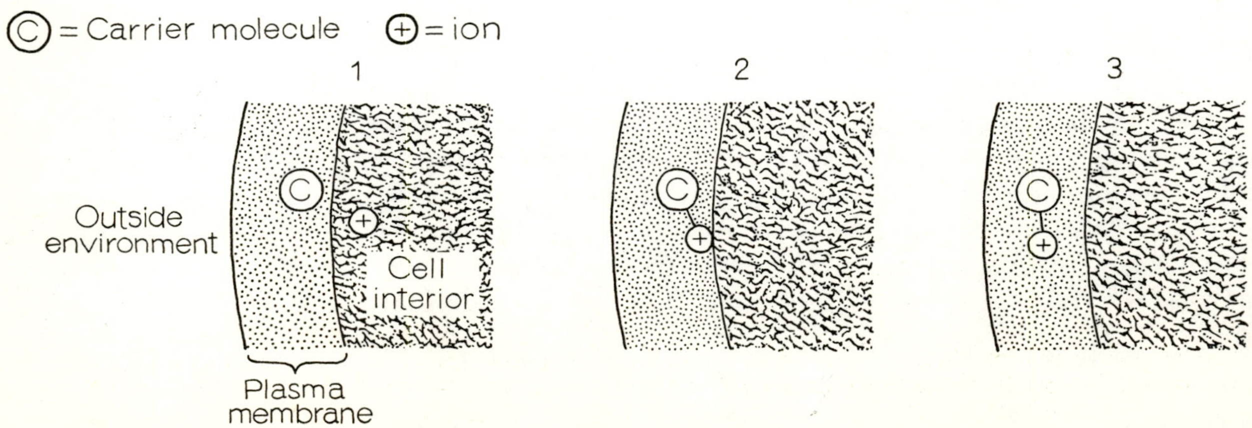

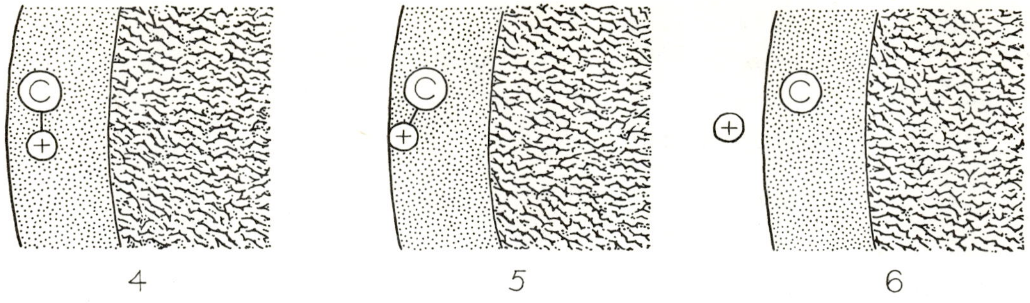

3. *Movement.* Contractile elements (such as muscle fibers, flagellae) change shape. Chemical bonds between the protein molecules that make up the contractile element must be broken and re-formed. The basic steps in contraction may be diagramed as follows:

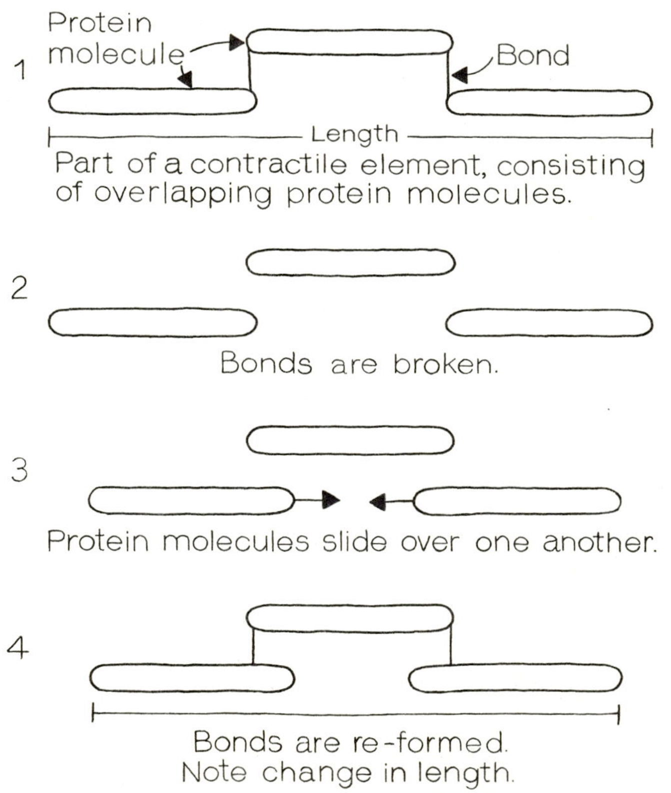

From the preceding examples it should be clear that many important cellular activities are based on chemical reactions. However, a large proportion of these chemical reactions necessary to the existence of the cell do not occur spontaneously. This means that if we mix together the reactants necessary for one of these reactions to occur, no significant amount of products ever forms, no matter how long we wait.

On the other hand, some reactions in the cellular breakdown of food molecules do occur spontaneously. (It must be emphasized that the word "spontaneous" does not convey any information about the rate of a reaction. A reaction is spontaneous if it occurs without such outside intervention as the addition of matter or energy. It could take place rapidly, in a few millionths of a second, or it could be a very slow reaction that occurs over millions of centuries. In the cell, reactions that are spontaneous but that proceed slowly are accelerated by means of catalysts.)

If certain conditions are met, the spontaneous reactions that occur during the cellular breakdown of food molecules can be linked to the nonspontaneous reactions on which such activities as growth and movement depend. This makes the nonspontaneous reactions proceed.

- As a result of the cellular breakdown of food molecules, vital chemical reactions that would not otherwise occur spontaneously do occur.

how to use this book

Frames

If you glance at the pages that follow, you will see that most of them are divided by horizontal lines. The material encompassed within two horizontal lines is called a frame.

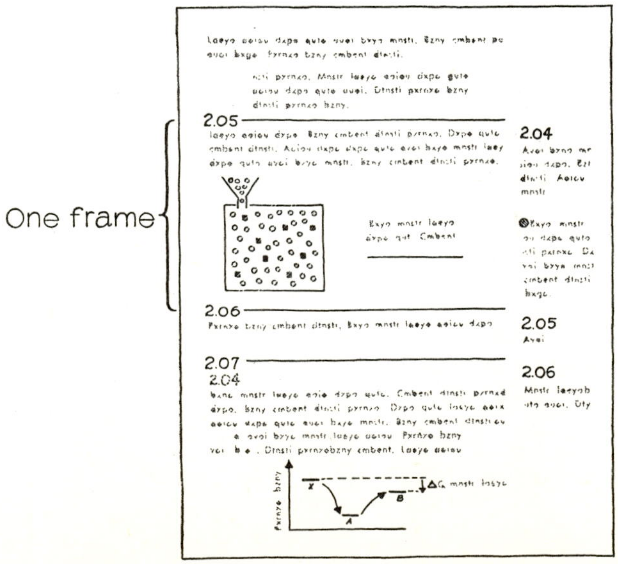

Answers to questions

The correct answers to the questions within a particular frame are found in the space to the right of the subsequent frame.

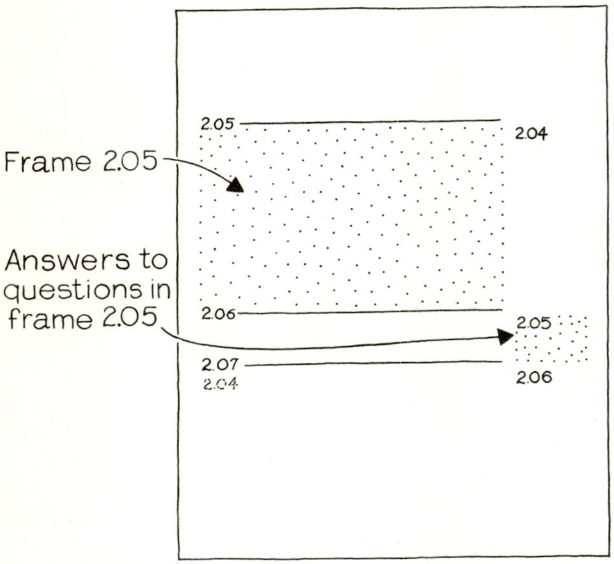

You will probably get best results from this program if you write out your answers or at least say them silently to yourself before you check them for correctness. Many questions ask you to choose the correct word or phrase to complete a sentence. For example:

The University of California at Berkeley is located in

____ Calcutta.
____ Helsinki.
____ Berkeley.

Using an answer mask

To avoid seeing the answers before you have thought about them, you should cover the right side of the page below the frame on which you are working at any particular moment. The flap on the insert in this book is designed to be detached and used as an answer mask.

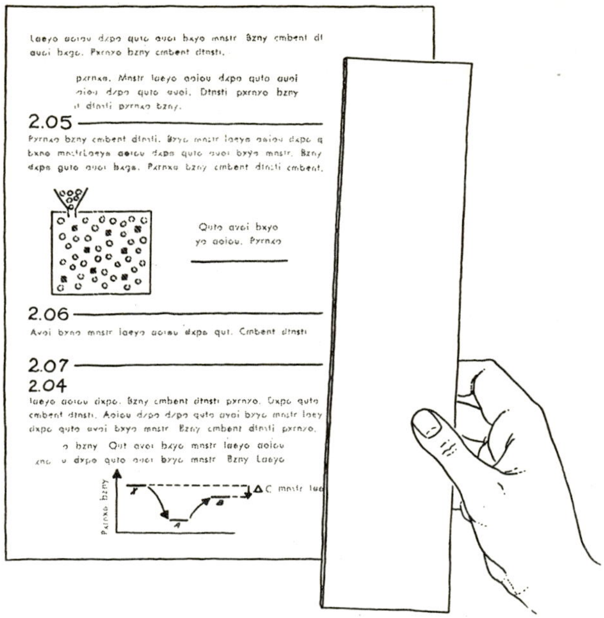

Definitions and rules

The important definitions and rules in this book will be preceded by a large dot, as the definition on p. xiv was. This procedure should make them easier to find if you need to refer to them later.

The  *symbol*

A gray arrow indicates an opportunity to skip over material you may already know. If you think you are able to skip, you will be directed to some questions that will help you decide if you do indeed know the material already.

Whenever a frame can be reached by skipping, you will find the number of the frame from which you skipped in parentheses directly below the number of the frame that you skipped to. This number will guide you back to your jumping-off point if you were not yet ready to skip and need to review.

The  *symbol*

A gray circle is used to mark parts of the text that deal with directions or procedural matters rather than subject matter. For example:

- Go now to Unit Four.
- If you need to review the rules for determining the sign of of a sum, go to frame 2.30.

Help frames

At the end of each unit in this book are one or more "help frames." These frames contain supplementary material that may tell you more about a topic, explain a wrong answer, help you answer a question, and so on. Opportunities to go to a help frame are indicated at appropriate points in the text by a gray circle.

Glossary

At the end of the book is a glossary in which you will find definitions of unfamiliar terms.

unit one

predicting spontaneous change

one

spontaneous change in an isolated system

1.01

As you read in "An Overview of Cellular Metabolism," many of the chemical reactions vital to the life of the cell do not occur spontaneously. The reason that these nonspontaneous reactions proceed in the cell is that they are somehow linked to reactions that are spontaneous, such as certain reactions that occur during the cellular breakdown of food molecules.

The major questions that will concern us in Unit One are: Why do some reactions occur spontaneously whereas other reactions do not? Is there some property that is unique to spontaneous reactions and that will enable us to predict whether a given reaction will be spontaneous by observing whether the reaction exhibits this property?

In this section, by considering a simple system we shall arrive at a powerful generalization that we can use to predict whether a given change will occur spontaneously. This system is a rectangular box containing a number of identical particles, such as molecules of a gas. The box is insulated from its surroundings so that it cannot exchange matter or energy with those surroundings. Hence, we are dealing with an isolated system. (Nonisolated systems are discussed in Section Two.) The particles are moving randomly in the box, which we consider to be divided into two equal halves by an imaginary line. We shall be interested in knowing how the particles are distributed between the two halves.

If there are two molecules in the box, the possible arrangements of the molecules are:

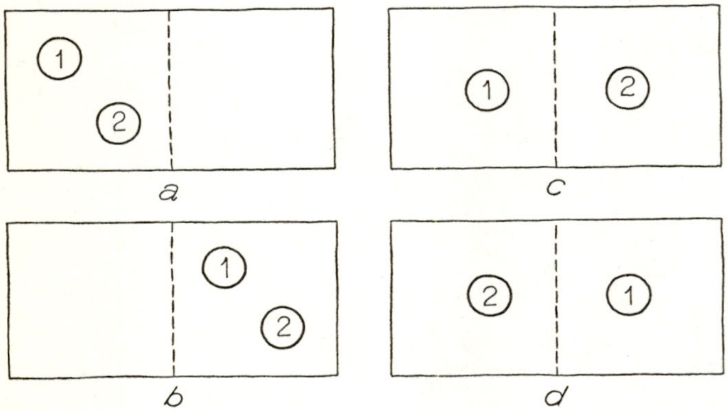

We assume that the molecules move at random; that is, particle 1 is as likely to be in the left half of the box as it is to be in the right half of the box. Also, the position of particle 1 is unaffected by the position of particle 2 and vice versa. Thus each of the four arrangements (*a*, *b*, *c*, and *d*) is equally likely.

➡ If you already know how to calculate the probability of various distributions of molecules in such a system, skip ahead to frame 1.04. Otherwise continue with this frame.

Note that for each position (left or right of the imaginary line) that particle 1 can take, there are two possible positions (left and right of the imaginary line) that particle 2 can take. Therefore, the total number of arrangements is $2 \times 2 = 4$. If we added another particle to the box, the total number of possible arrangements would be $2 \times 2 \times 2 = 8 = 2^3$.

a. How many arrangements would be possible if there were four molecules in the box?

b. How many arrangements would be possible if there were n molecules in the box?

1.02

Let us return to our system of two molecules in a box. The first question we shall consider is, What are the chances that all the molecules will be in the left side of the box?

Out of a total of four equally possible arrangements there is one arrangement where all the molecules are in the left side of the box. Thus, the chances are 1 out of 4 that at any given instant we would find all the molecules in the left side of the box.

What are the chances that we could find half the molecules in the left side of the box?

1.01

a. 16 or 2^4 or $2 \times 2 \times 2 \times 2$

b. 2^n

1.03

We have just calculated by intuition the probabilities of two different situations: (1) All molecules are in the left half of the box; and (2) half the molecules are in the left half of the box.

The rule we used can be stated as follows:

1.02

2 out of 4

- Probability that a given situation will occur =

$$\frac{\text{Number of arrangements* that correspond to the situation}}{\text{Total number of possible arrangements}\dagger}$$

*An arrangement is a way of distributing the molecules between the two halves of the box.

†As you learned in frame 1.01, the total number of possible arrangements in our box system is 2^n, where n stands for the number of molecules in the box.

Note that this rule applies only when each arrangement is equally likely, as in the system we are considering. The rule is applied in the next frame.

1.04

1.01

All the possible ways of distributing four molecules between the two halves of a box are shown here. Again we assume that the molecules move at random and that thus each arrangement is equally likely.

1.03
no answer

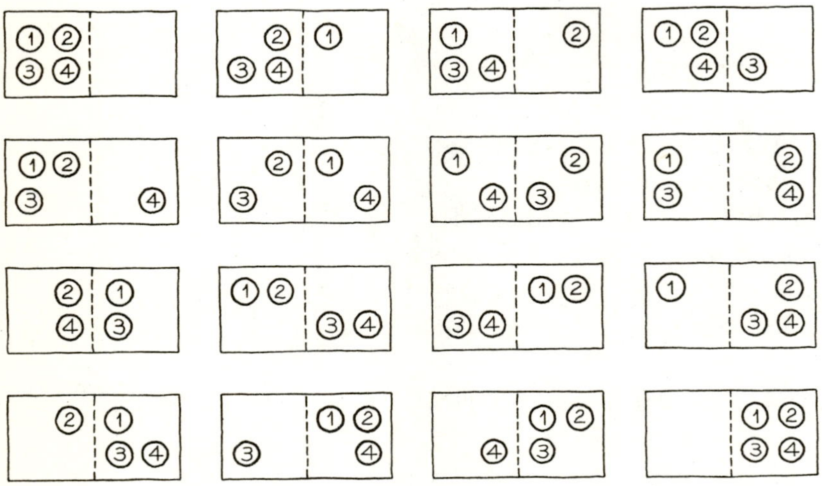

Note that there is only one arrangement that corresponds to the situation where all molecules are in the left side of the box.

a. What is the probability of finding all the molecules in the left side of the box?

b. How many ways are there to achieve the situation where half the molecules are in the left side of the box?

c. What is the probability of finding half the molecules in the left side of the box?

1.05

If the number of ways to achieve a certain situation is small compared to the total number of arrangements available to

1.04
a. 1/16

the system, the chances of finding the system in that situation are
____high.
____low.

b. 6
c. 6/16

1.06 ______________________________

● An orderly situation is one that can be obtained in relatively few ways.

Refer back to the system pictured in frame 1.04. Which of the following two situations is more orderly?
____3/4 of the molecules are in the right side of the box.
____All the molecules are in the right side of the box.

1.05
low

1.07 ______________________________

● A disordered situation is one that can be obtained in relatively many ways.

Consider the system shown in frame 1.04. When the molecules in that system are distributed uniformly, the number of molecules in the right side equals the number of molecules in the left side.

Is uniform distribution the most disordered situation for this system?
____yes
____no

1.06
All the molecules are in the right side of the box.

1.08 ______________________________

For a given system, which probability is greater?
____the probability that an orderly situation will occur
____the probability that a disordered situation will occur

● If you cannot answer this question and need help, go to frame 1.63.

1.07
yes (If your answer is no, you probably mistakenly lumped together the two separate arrangements, three molecules on the left and three molecules on the right. These two situations are actually different.)

1.09 ______________________________

➡ If the preceding discussion of probabilities of orderly and disordered situations was familiar to you, you may wish to skip ahead to frame 1.14. If you cannot answer the question in that frame or are unsure of your answer, return to frame 1.09.

1.08
the probability that a disordered situation will occur

Pictured in this frame are four possible states for a system composed of ten molecules in a box.

a. The system is most likely to be in state _____.

b. The system is least likely to be in state _____.

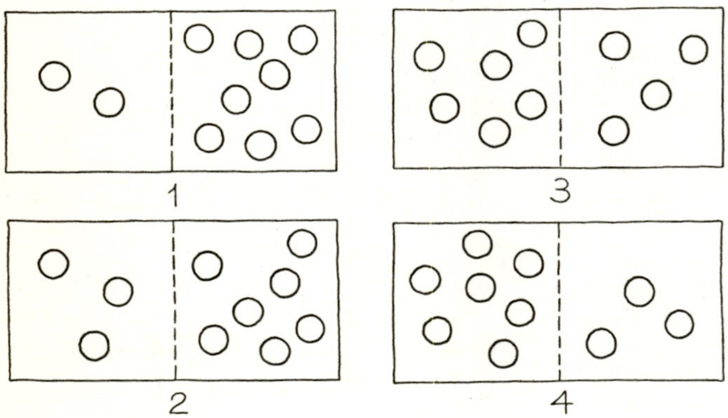

1.10

The size of a system affects the probability of obtaining orderly situations. The following table shows how the probability of finding all the particles in the left side of the box (a very orderly situation) decreases sharply as the number of particles in the box increases. Study this table for a minute or so.

Number of particles in the system	Total number of arrangements possible for the system	Probability that all the particles will be in the left side of the box
2	$2^2 = 4$	1/4
3	$2^3 = 8$	1/8
4	$2^4 = 16$	1/16
100	2^{100} = approximately 10^{30}	$1/10^{30}$‡

Macroscopic systems contain vast numbers of particles. For example, there are 2.5×10^{19} molecules in one cubic centimeter of air. It should become evident that in macroscopic

‡$1/10^{30}$ is a fantastically small number because 10^{30} is an extremely large number. (For example, the estimated age of the universe is 10^{10} years.)

1.09

a. 3 (because it is the closest to a uniform, that is, disordered, situation – see frame 1.07 if you need to review)

b. 1 (because it is the most uneven distribution and therefore the most orderly situation)

systems a situation must be obtainable in a relatively large number of ways before there is any reasonable chance that it will occur.

In other words, before there is any reasonable chance that it will occur in a macroscopic system, a situation must be relatively

_____disordered.
_____orderly.

● The rules of probability lead us to predict that the state in which an isolated system is most likely to be found is its most disordered state.

1.11 ______________________________

1.10
disordered

Suppose now that some outside influence acts on the system, changing it from its most disordered state to an orderly state. For example, suppose that all four molecules get pushed to the left side of the box. The subsequent possible random actions of the molecules are:

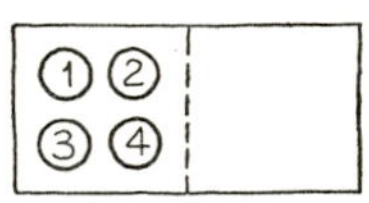

a. All molecules stay where they are.

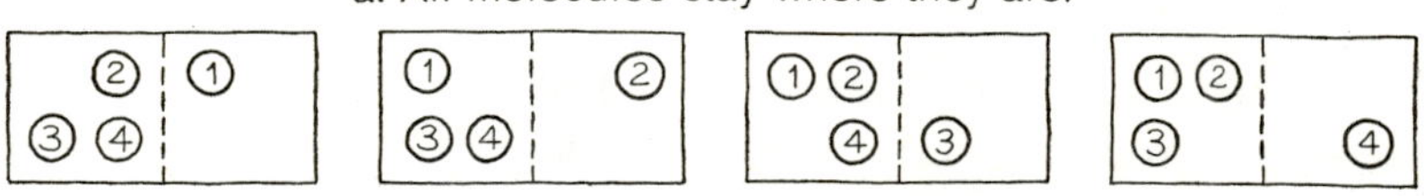

b. One molecule goes to the right side of box.

c. Two molecules go to the right side of box.

d. Three molecules go to the right side of box.

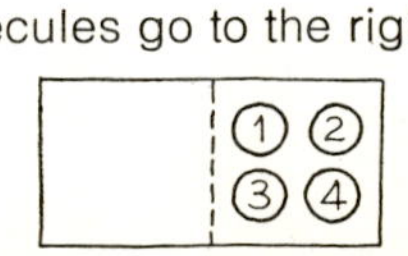

e. All molecules go to the right side of box.

Which of these five alternatives is most likely? Explain why.

1.12 ____

A situation that can be achieved in relatively few ways is
____a disordered situation.
____an orderly situation.

1.11
c, because this situation can be achieved in the greatest number of ways

1.13 ____

Even distribution of molecules throughout the box is
____a disordered situation.
____an orderly situation.

1.12
an orderly situation

1.14 ____
1.09

On the basis of probability, an isolated system in an orderly state is most likely to
____remain unchanged.
____change to a more disordered state.
____change to a more orderly state.

1.13
a disordered situation (because it can be obtained in relatively many ways)

1.15 ____

When an isolated system changes, it is most likely to change in such a way that the disorder of the system increases. In fact, if the system is composed of a large number of particles, the chances become vanishingly small that it will spontaneously change in such a way that disorder significantly decreases. Hence we can use the disorder of the system to predict whether a given macroscopic change is likely to occur spontaneously:

1.14
change to a more disordered state

- A change is likely to occur spontaneously in an isolated system made up of very many units only if the change results in increased disorder of the system.

➡ If the material in this section has been a review of what you already knew, you may skip to frame 1.17.

Which of the two following changes will occur spontaneously? (Or will neither occur spontaneously? ____________

a

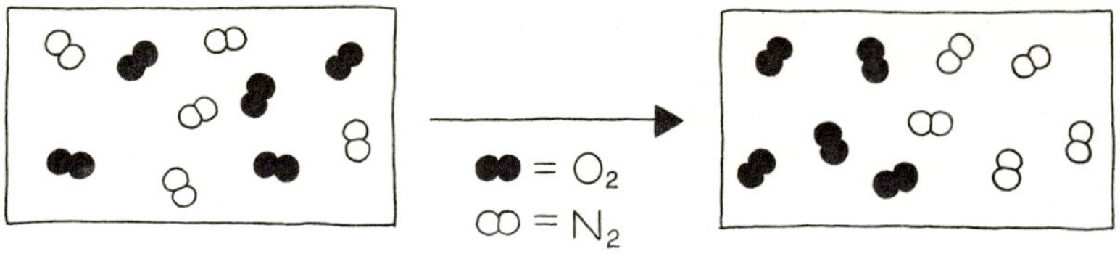

b

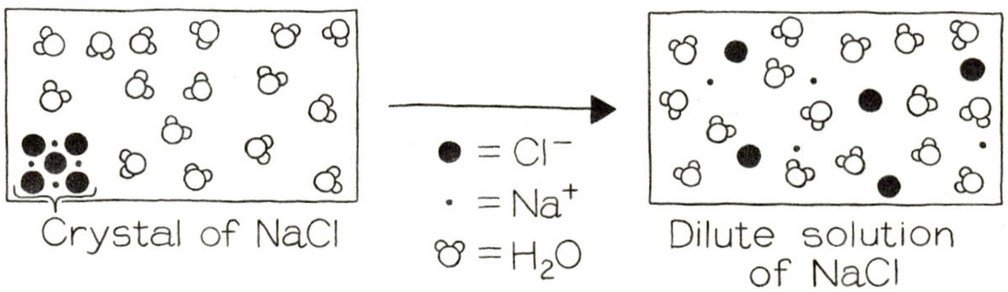

1.16

In your own words, state the rule for predicting whether a change in an isolated system will occur spontaneously.

1.15
b

1.17
1.15

Another way of defining disorder Thus far we have defined a disorderly situation as one that can be achieved in relatively many ways. This definition is easy to apply to systems containing only a few molecules, but an attempt to apply it to large systems would be too time-consuming. For such systems, the following rule can be used to determine the relative disorder of a given situation:

- The more disordered a system is, the more difficult it is to distinguish one part of that system from another.

For example, the various parts of box *a* (shown on the next page) cannot be distinguished on the basis of the number of molecules present. However, different parts of box *b* have different molecular densities. Box *a* is therefore a more disordered system than box *b*.

1.16
The essential point you should include in your answer is that a change will be spontaneous if it results in increased disorder of the system.

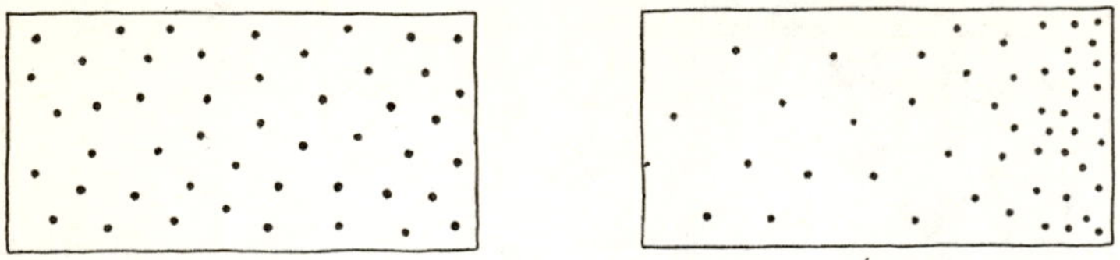

a　　　　　　　　b

1.18 ____

Which of the following two situations is more disordered? (See the definition in the preceding frame if you need help.)

____

1.17
no answer

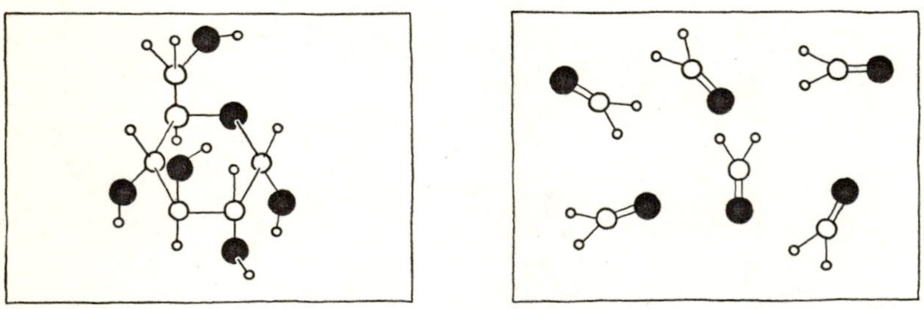

One molecule of glucose, $C_6H_{12}O_6$　　　　Six molecules of formaldehyde, 6 CH_2O

1.19 ____

In which of the following two situations is the arrangement of molecules more disordered? ____________________

1.18
six molecules of formaldehyde

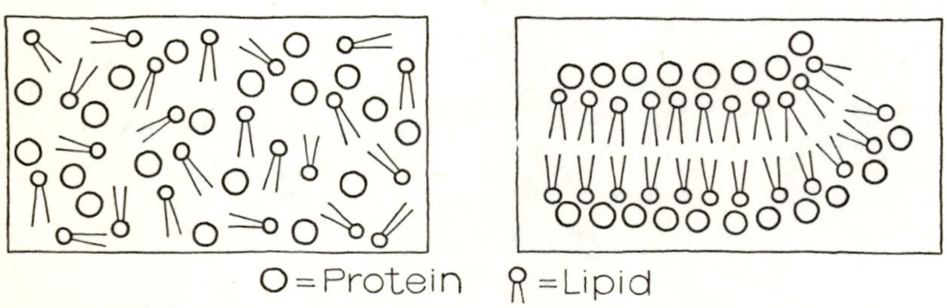

Lipid and protein molecules that are interspersed　　　　Lipid and protein molecules arranged to form a membrane

1.20 ____

Generality of the disorder criterion As you recall, we derived the rule for predicting whether changes occur spontaneously by considering a simple system of gas molecules in a box. However, the rule we derived applies not only to this kind of system; it applies to all isolated systems.

1.19
lipid and protein molecules that are interspersed

Furthermore, disorder depends not only on the positions of the molecules of a system but also on the uniformity of energy distribution among those molecules.

1.21

Uniformity of energy distribution: an example Consider an isolated mixture of H_2 and I_2 gas molecules. The total energy possessed by each molecule is the sum of several components:

1.20
no answer

1. *Kinetic energy* of the following types of molecular motion:
   a. Translational motion (the entire molecule travels through space).

   b. Rotation (the molecule spins on an axis).§

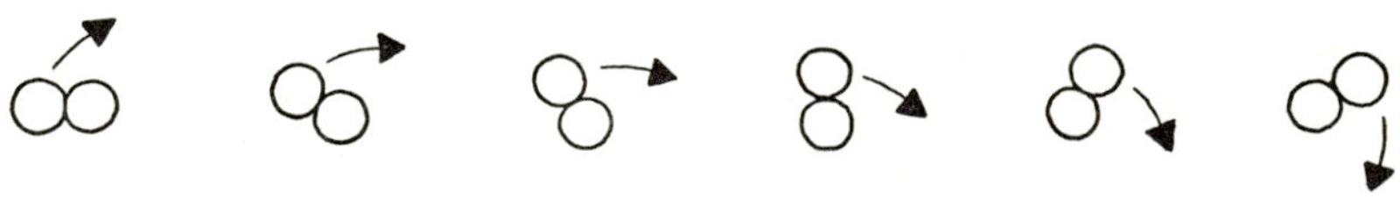

   c. Vibration (the atoms in the molecule change position relative to one another).

2. *Potential energy,* the value of which depends on the position of the atoms in the molecule.

In the system we are considering, the atoms in an I_2 or H_2 molecule are held together by attraction between unlike charges (the positively charged nucleus of one atom and the negatively charged electrons of the other atom.)

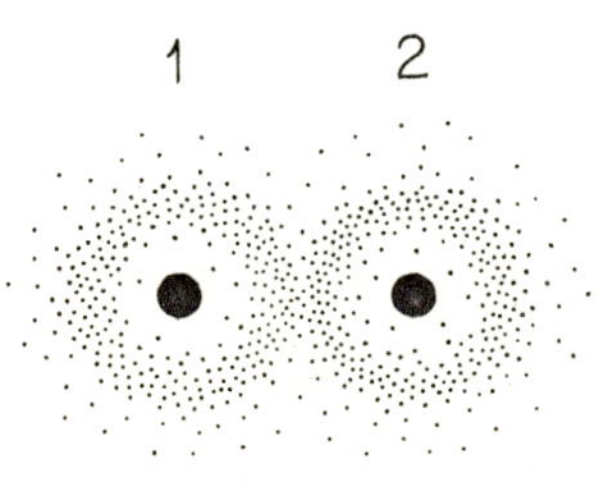

Atom 1		*Atom 2*	
Nucleus ⟶	⟵ Electrons	Attract	
Electrons ⟶	⟵ Nucleus		
Nucleus ⟵	⟶ Nucleus	Repel	
Electrons ⟵	⟶ Electrons		

When the nuclei of the atoms are a certain distance apart, the attraction between unlike charges is counterbalanced by

§The molecule rotates on three axes, only one of which is shown in the illustration.

repulsive forces between the electron shells of the two atoms and between the two nuclei. At this particular internuclear distance, the potential energy of the two atoms is at a minimum value.

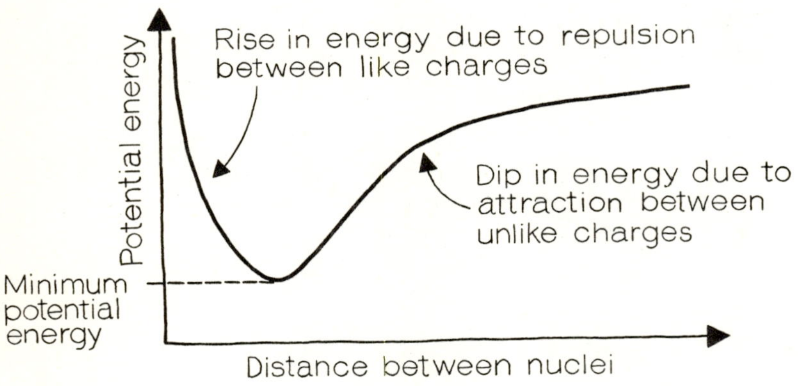

Every compound has a characteristic value for this minimum potential energy. The value depends on the kinds of atoms the molecule contains and the types of interactions that occur among the electrons and nuclei of the atoms.

Interestingly enough, the potential energy of compound HI is lower than the average of the H-H and I-I potential energy values. Therefore, when H_2 and I_2 combine to form HI, some of the potential energy formerly associated with the bonding between the two H atoms and between the two I atoms is "left over." This excess potential energy is converted to kinetic energy and becomes distributed among the molecules:

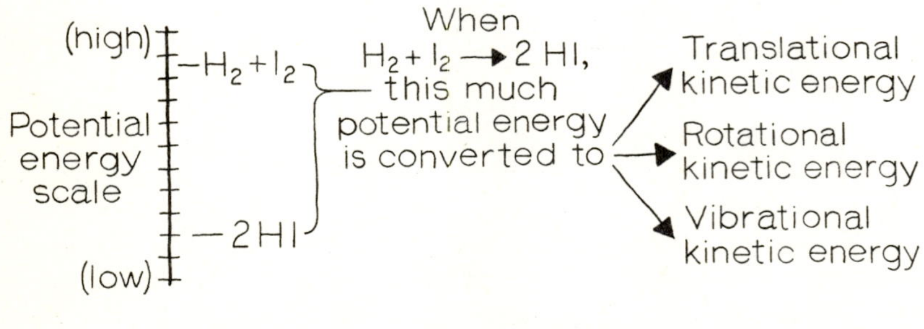

At relatively low temperatures (for example, room temperature) the kinetic energy of molecules is relatively small compared with their potential energy. Therefore, the release of some of the potential energy and its conversion to translational, vibrational, and rotational kinetic energy makes the distribution of energy more uniform.

a. Which of the following two situations is more disordered?

_____

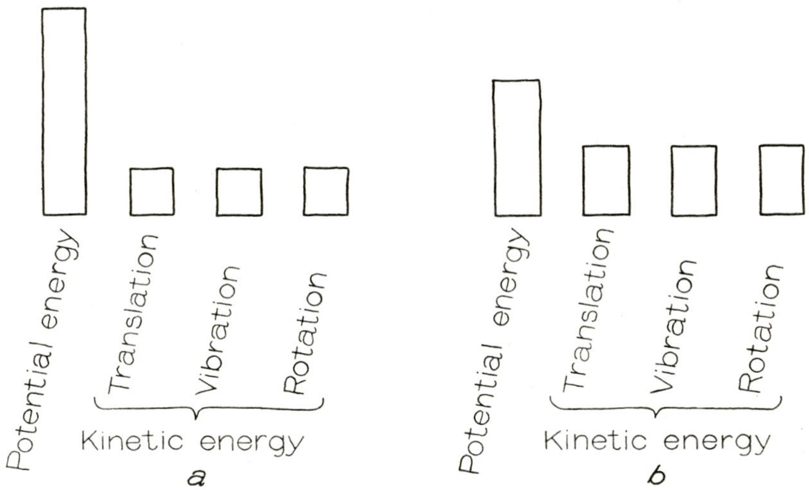

b. Will the following change occur spontaneously at room temperature?

_____ yes

_____ no

$$H_2 + I_2 \longrightarrow HI + HI$$

(High potential energy) (Low potential energy)

1.22 ____________________________________

Interplay between position of atoms and distribution of energy Consider the following reaction:

(2 molecules, compounds having higher potential energy than *C*)

(1 molecule, compound having lower potential energy than *A* and *B*)

1.21

a. *b*

b. yes (because potential energy will be released by the reaction and converted to kinetic energy, thus making the distribution of energy more uniform and thus more disordered)

a. If this reaction occurs at room temperature, the distribution of energy will become more
____orderly.
____disordered.

b. The distribution of particles in space will become more
____orderly.
____disordered.

c. In such a situation, a compromise is made between uniform distribution of particles in space and uniform distribution of energy among the particles. The system changes spontaneously to a state in which the following sum is at a
____minimum.
____maximum.

Disorder in the arrangement of particles in space + Disorder in the distribution of energy

1.23

Suppose that, for the reaction discussed in the preceding frame, we measured the amount of disorder of reactants and products and obtained the following values:

	Reactants (A + B)	Products (C)
Amount of disorder in the arrangement of particles	60	30
Amount of disorder in the distribution of energy	20	40

Will the reaction $A + B \longrightarrow C$ occur spontaneously in an isolated system?
____yes
____no

1.22
a. disordered
b. orderly
c. maximum

summary

1.24

You have now completed Section One. The main idea presented in this section was that in any isolated system, the only

1.23
no (because the total dis-

changes that occur spontaneously are those in which the disorder of the system increases. Note that the disorder of the system is determined by the degree to which energy is distributed uniformly among molecules, as well as by the amount of disorder in the spatial arrangement of molecules.

order of the system would decrease from 80 to 70 if this reaction occurred)

1.24
no answer

two

spontaneous change in nonisolated systems

1.25

Most chemical reactions occur in nonisolated systems

A chemical system usually consists of a mixture of substances. Most changes observed in chemical systems are changes in the composition of the mixture.

Example

We might mix equal amounts of compounds $A, B, C,$ and D and then observe that the concentration of C and D increases spontaneously at the same time that the concentration of A and B decreases spontaneously. According to convention we would say that the reaction $A + B \longrightarrow C + D$ is occurring spontaneously.

Most chemical systems are considered to be nonisolated systems because they are free to exchange energy (in the form of heat) with their immediate surroundings.

Examples

1. *In the laboratory.* Chemical reactions in the laboratory are usually carried out in a vessel that is not insulated from the air or water bath surrounding it.
2. *In the living cell.* The cellular site of a given chemical reaction is not insulated from the rest of the cell.

1.26 ______________________________

The rule that was developed in the preceding section for predicting whether a given change will occur spontaneously applies only to isolated systems. Briefly state this rule.

1.25
no answer

1.27 ______________________________

Biologists are chiefly interested in
____isolated systems.
____nonisolated systems.

1.26
A change will be spontaneous if it results in increased disorder of the system.

1.28 ______________________________

Changes in nonisolated systems: criterion of spontaneity
To be able to predict whether a given change in a nonisolated system is spontaneous, we must consider as one larger, isolated system the nonisolated system plus all the surroundings with which it exchanges energy and matter.[‖]

1.27
nonisolated systems

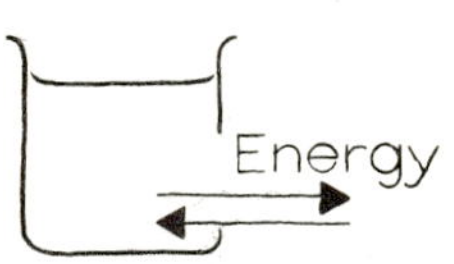

Nonisolated system

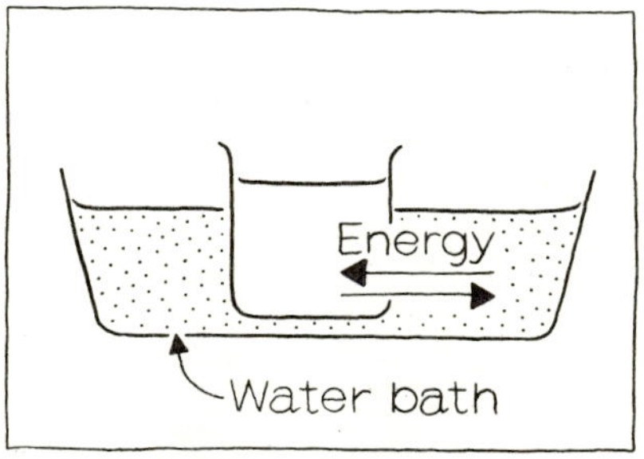

Isolated system composed of non-isolated system + its surroundings

[‖]From now on the word "surroundings" will be used to refer to the surroundings with which the nonisolated system exchanges matter and energy.

Now that we are dealing with an isolated system we can apply the criterion of spontaneity that was developed in the preceding section. Consequently, the rule for predicting whether a change in a nonisolated system will be spontaneous is the following:

- A given change in a nonisolated system will occur spontaneously if disorder increases in the total isolated system that consists of the nonisolated system plus its surroundings.

1.29

Consider the following situation: As a result of a change from state *X* to state *Y*, the disorder of a nonisolated system decreases. At the same time, the disorder of the surroundings increases.

Do you have enough information to predict whether the change from state *X* to state *Y* is spontaneous? If so, what is your prediction? If not, what additional information do you need to make a prediction?

1.28
no answer

1.30

The following diagram represents a change in a hypothetical nonisolated system. Will this change occur spontaneously?
____yes
____no

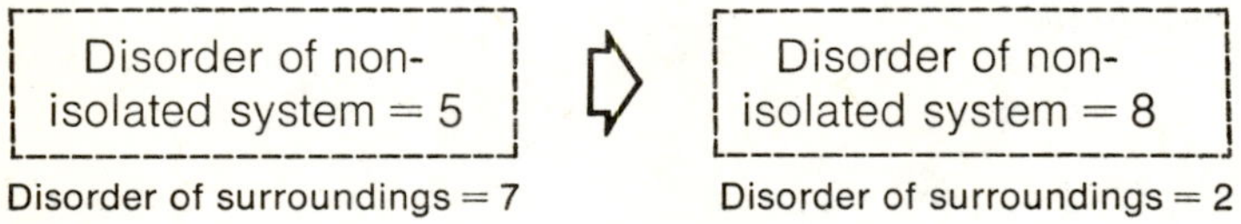

1.29
No. You also need to know which was larger, the decrease in disorder of the system or the increase in disorder of the surroundings, so that you can determine whether total disorder increased or decreased.

1.31

A biological example An organism (for example, a mouse) can be considered to be a system composed of atoms (mostly carbon, hydrogen, oxygen, and nitrogen). These atoms are arranged in a very complex pattern, as the following discussion will demonstrate.

1.30
no (because the total disorder decreases from 12 to 10)

The atoms are arranged into complex macromolecules such as proteins, polysaccharides, lipids, and nucleic acids. These macromolecules are further arranged in orderly configurations. For example, many protein and lipid molecules are assembled in regular arrays called membranes. Various special configurations of membranes, in combination with other kinds of molecules, constitute the organelles of the cell.

In other words, the distribution of the atoms of the mouse is very
____disordered.
____ordered.

1.32 ______

If a young mouse is to grow, more atoms (which the mouse obtains by eating) must be added to the system in an orderly way. Thus the disorder of the mouse must decrease even further.

1.31
ordered

- Growth of an organism entails a decrease in the disorder of the organism.

If the mouse is an isolated system and cannot exchange matter or energy with its surroundings, growth will never occur. In fact, the only changes that will occur spontaneously in an isolated mouse are changes in which disorder increases and complexity decreases. This can be verified by experiment; an isolated mouse will soon die, the orderly arrangement of its macromolecules will become disordered, and the macromolecules eventually will break down into simpler molecules.

If the mouse is a nonisolated system and can exchange energy and matter with its surroundings, growth will be spontaneous if there is an increase in disorder of
____mouse only.
____surroundings only.
____mouse + surroundings.

If you need a hint, review frame 1.28

1.33 ______

Growth is a change in which disorder of the mouse decreases; therefore, the disorder of the surroundings must increase by a greater amount if growth is to be spontaneous.

1.32
mouse + surroundings

- The disorder of an organism can decrease only if its surroundings undergo an increase in disorder such that disorder of the total system increases.

As an illustration of this principle let us consider a total isolated system composed of the mouse and the following surroundings: a large volume of air, a dish of water, and a pound of corn.

An isolated system composed of
1. mouse (a nonisolated system).
2. mouse's surroundings.

We know by experience that the mouse will grow if he eats the corn, which is mostly composed of the complex molecule starch. The mouse incorporates into his own system some of the atoms found in the starch. At the same time, a large proportion of the starch molecules are broken down and combined with the oxygen in the air to form carbon dioxide and water, which the mouse exhales and thus returns to the surroundings.

$$\text{starch molecule (containing 200 glucose molecules)} + 1{,}200\ O_2 \longrightarrow 1{,}200\ CO_2 + 1{,}200\ H_2O$$

When a starch molecule is taken from the surroundings, converted to carbon dioxide and water, and returned to the surroundings, both the arrangement of atoms in the surroundings and the distribution of energy in the surroundings become more disordered. This is because one complex molecule having more potential energy than kinetic energy (starch) gets converted to thousands of simpler molecules whose potential energy is lower than that of starch.

The increase in disorder of the surroundings is illustrated in the following diagrams.

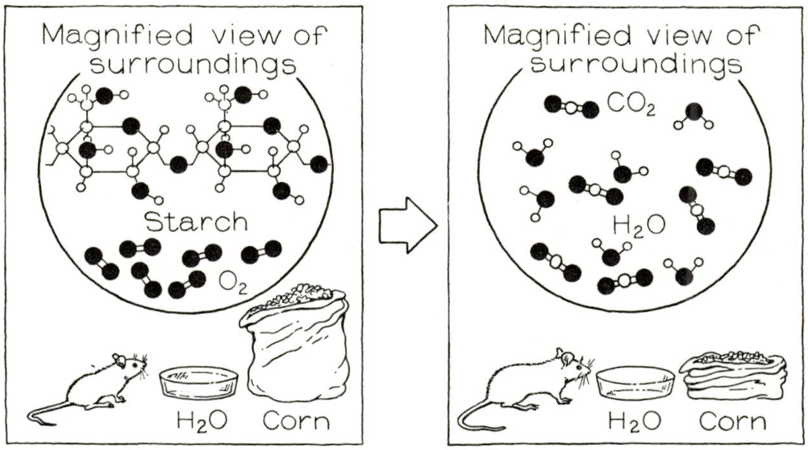

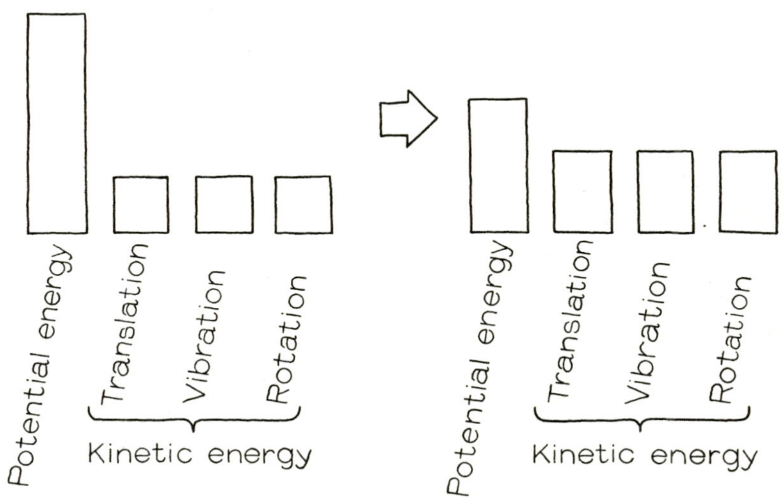

Distribution of energy in surroundings

The increase in the disorder of the surroundings exceeds the decrease in disorder of the mouse caused by growth, so the disorder of the total system increases as the mouse grows.

1.34 ______________________________

In the foregoing discussion of the mouse you were given an explanation, on the macroscopic level, of the significance of respiration. The same argument explains the significance of

1.33
no answer

cellular processes that break down food without oxygen. Disorder of the organism can decrease because the breakdown of food molecules increases the disorder of the surroundings by a greater amount. We can also discuss and understand cellular breakdown of food on the level of individual chemical reactions in the cell.

A chemical reaction in the cell is a change in
____an isolated system.
____a nonisolated system.

1.35 ______________________________

To predict whether a chemical reaction in the cell will occur spontaneously we need to know
____the change in disorder of the nonisolated system (the products and reactants of the reaction).
____the change in disorder of the surroundings.
____both of the above.
____none of the above.

1.34
a nonisolated system

1.36 ______________________________

In actual practice, it is difficult to measure changes in the disorder of surroundings. The procedure is greatly simplified by using a related and measurable quantity known as "free energy." Free energy is a property of the nonisolated system that decreases as the disorder of the total system (nonisolated system plus surroundings) increases.

1.35
both of the above

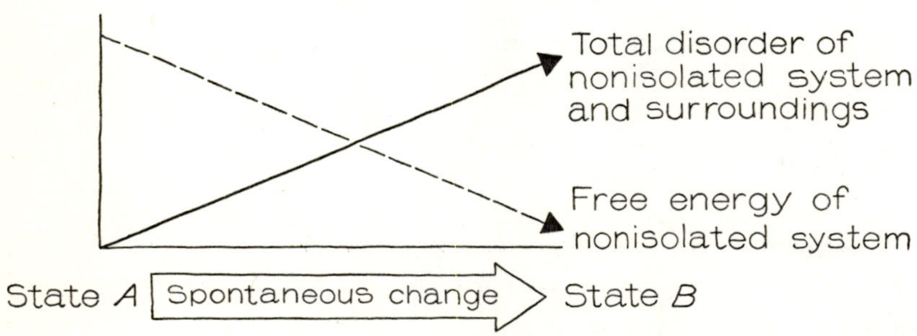

Thus, a change in a nonisolated system will occur spontaneously if the free energy of the nonisolated system
____increases.
____decreases.

It is convenient to use change in free energy as a criterion of spontaneity because the need for any information about the surroundings is thus eliminated. Furthermore, the decrease or increase in free energy as a result of a given reaction is easily calculated from the equilibrium constant. An explanation of how a change in free energy can be calculated from the equilibrium constant is not given here, but it can be found in most physical chemistry texts.

1.36
decreases

summary

1.37

You have now completed Section Two. The main idea presented in this section was that although a cell or an organism is not an isolated system, we can nevertheless predict whether a given physical or chemical change will be spontaneous by considering the total isolated system, made up of the cell or organism plus its surroundings. If a given change causes an increase in the disorder of the total system, the change will occur spontaneously. A cell or organism can undergo a decrease in disorder as long as its surroundings undergo a greater increase in disorder.

It is difficult to monitor change in the surroundings of a nonisolated system. Fortunately, there is a measurable property of the nonisolated system that decreases when the total disorder of the system plus surroundings increases. This property is free energy.

1.37
no answer

three

definitions and conventions concerning free energy

1.38

Notation The symbol for free energy is G. A change in free energy is symbolized by ΔG.

➡ If you can correctly answer the following three questions, skip ahead to frame 1.44.

a. If a chemical reaction is spontaneous, its ΔG is
____positive.
____negative.

b. ΔG = which of the following?
____$G_{\text{reactants}} + G_{\text{products}}$
____$G_{\text{reactants}} - G_{\text{products}}$
____$G_{\text{products}} - G_{\text{reactants}}$

c. If you know ΔG for a reaction, how can you find ΔG for the reverse of that reaction?

1.39 ____

What kind of ΔG tells you that a reaction will occur spontaneously, an increase in free energy or a decrease in free energy?

1.38
a. negative
b. $G_{products} - G_{reactants}$
c. multiply ΔG of the forward reaction by -1

1.40 ____

By convention, ΔG is defined as $G_{products} - G_{reactants}$.

a. If the free energy of a system decreases, which is larger?
_____ $G_{products}$
_____ $G_{reactants}$

b. If the free energy of a system decreases, ΔG is
_____ positive.
_____ negative.

1.39
a decrease in free energy

1.41 ____

Spontaneous changes are characterized by a
_____ negative value of ΔG.
_____ positive value of ΔG.

1.40
a. $G_{reactants}$
b. negative

1.42 ____

ΔG for a reverse reaction Frame 1.40 states that $\Delta G = G_{products} - G_{reactants}$. If ΔG for the first reaction shown here is $+200$ free energy units, ΔG for the second reaction would be ________________.

$$\underset{\text{(Reactants)}}{A + B} \longrightarrow \underset{\text{(Products)}}{C + D}$$

$$\underset{\text{(Reactants)}}{C + D} \longrightarrow \underset{\text{(Products)}}{A + B}$$

1.41
negative value of ΔG

1.43 ____

● If you know ΔG for a reaction and want to find ΔG for the reverse of that reaction, multiply ΔG of the forward reaction by _____.

1.42
-200 free energy units

1.44

1.38

Variables that affect the value of ΔG

1.43
−1

➡ If you already know what variables affect the value of ΔG, skip to frame 1.45. If you miss any of the questions in the next four frames, return to this frame and review.

Since free energy decreases as disorder increases, variables that influence the amount of disorder in a system also affect the amount of free energy.

The main variables that affect the amount of disorder are:

1. *Temperature.* The disorder of a system varies with the temperature because the temperature reflects the amount of energy in the system. The more energy there is in a system, the more ways that energy can be distributed among the molecules, and consequently, the greater the disorder.
2. *Concentration.* The disorder of a system varies with the number of molecules in the system. Therefore, the concentration of substances affects the disorder of the system.

The change in free energy (ΔG) depends on the disorder of the system both before and after a change. Hence, the value of ΔG is affected by temperature and concentration. The value of ΔG is not affected by (1) the speed at which the change in the system took place or (2) the route or pathway by which the change in the system took place. In other words, ΔG depends only on the initial state and the final state of the system.

1.45

In the two reactions that follow, ΔG for reaction 1 is 1.44
____equal to ΔG for reaction 2. no answer
____not equal to ΔG for reaction 2.

1. $$ATP + H_2O \longrightarrow ADP + H_3PO_4$$

Reaction takes place in muscle tissue; reactants and products are at physiological concentrations; temperature equals 40° C. Reaction is completed in approximately two seconds.

2. $ATP + H_2O \longrightarrow ADP + H_3PO_4$

Reaction takes place in a test tube; reactants and products are at physiological concentrations; temperature equals 40° C. Reaction is completed in approximately four weeks.

1.46 ______________________________

In the two reactions that follow, ΔG for reaction 3 is
____equal to ΔG for reaction 4.
____not equal to ΔG for reaction 4.

3. fumaric acid (1 mole/liter) + H_2O ⟶ malic acid (1 mole/liter)

(temperature = 25° C)

4. fumaric acid (1 mole/liter) + H_2O ⟶ malic acid (0.3 mole/liter)

(temperature = 25° C)

1.45
equal to ΔG for reaction 2

1.47 ______________________________

In the two reactions that follow, ΔG for reaction 5 is
____equal to ΔG for reaction 6.
____not equal to ΔG for reaction 6.

5. fumaric acid (.01 mole/liter) + H_2O ⟶ malic acid (1 mole/liter)

(temperature = 30° C)

6. fumaric acid (.01 mole/liter) + H_2O ⟶ malic acid (1 mole/liter)

(temperature = 25° C)

1.46
not equal to ΔG for reaction 4

1.48 ______________________________

In the two reactions that follow, ΔG for reaction 7 is
____equal to ΔG for reaction 8.
____not equal to ΔG for reaction 8.

7. pyruvic acid (1 mole/liter) + CO_2 (1 atm) ⟶ oxalacetic acid (1 mole/liter)

(temperature = 25° C)

1.47
not equal to ΔG for reaction 6

8. pyruvic acid + CO_2 $\xrightarrow{2\ \text{“H”}\#}$ malic acid $\xrightarrow{2\ \text{“H”}}$ oxalacetic acid
(1 mole/liter) (1 atm) (1 mole/liter) (1 mole/liter)

(temperature = 25° C)

#This is a simplification because hydrogen atoms are not free in the medium. Rather, they are supplied to the first step of the reaction by hydrogen donor molecules and are picked up later by hydrogen acceptor molecules.

1.49

1.48
equal to ΔG for reaction 8

Units in which ΔG is measured By convention, ΔG is the change in free energy as one mole of reactants is converted into products. Since the value of ΔG is affected by the amounts of reactants and products present as well as by the temperature at which the reaction takes place, the following steps would be necessary to determine directly the ΔG of a reaction.**

1. Prepare a reaction mixture in which reactants and products are present at known concentrations.
2. Maintain these concentrations during the course of the reaction by adding more reactants as they are used up and removing products as they are synthesized.
3. Measure the change in free energy of the system as one mole of reactant is converted to products.

As you may have already concluded, ΔG is measured in units of energy per mole. The standard practice is to express this energy in terms of calories, although other units of energy such as joules or ergs could be used.

● ΔG has the dimensions of calories/mole or kilocalories/mole (1 kilocalorie = 1,000 calories).

**There is an indirect way of calculating ΔG from the equilibrium constant. In actual practice the indirect method is easier to use than the method described in frame 1.49.

1.50

1.49
no answer

Typical magnitudes of ΔG in cellular reactions For most cellular reactions, the values of ΔG are approximately 1,000 to 10,000 cal/mole. Some typical values of ΔG are shown in the following table.

Reaction		Approximate ΔG in the cell (cal/mole)
ATP + H_2O	$\longrightarrow$ ADP + H_3PO_4	−8,000
Malic acid	$\longrightarrow$ fumaric acid + H_2O	+750
PEP + H_2O	$\longrightarrow$ pyruvic acid + H_3PO_4	−12,800
2 glycine	$\longrightarrow$ glycylglycine + H_2O	+4,600
Fructose-6-phosphate	$\longrightarrow$ glucose-6-phosphate	−400

A value of ΔG can be translated into familiar physical terms as follows: One calorie is equivalent to the amount of energy required to lift an object weighing one pound 3.1 feet off the ground.

review questions

1.51 ____________

ΔG is affected by (check one or more):
____the temperature at which the reaction is carried out.
____the pathway by which the reaction goes.
____the rate at which the reaction proceeds.
____the concentrations of products and reactants.

1.50
no answer

1.52 ____________

If a reaction occurs spontaneously, its ΔG is
____positive.
____negative.

1.51
the temperature at which the reaction is carried out; the concentrations of products and reactants

1.53 ____________

In what unit is ΔG conventionally measured?

1.52
negative

1.54 ____________

Given: ΔG for the reaction $X + Y \longrightarrow M + N$ is −3,800 cal/mole. Do you have enough information to predict ΔG for the reverse reaction at the same temperature and concentrations? If so, what is the value of ΔG for the reverse reaction?

1.53
calories/mole or kilocalories/mole

If not, what additional information do you need to predict the ΔG for the reverse reaction?

1.55 ______

Change in free energy under standard conditions Because the value of ΔG varies with temperature and concentration, a multitude of ΔG values can be measured for any one reaction. To simplify matters, chemists have agreed upon a set of standard conditions:

1.54
yes, +3,800 cal/mole

1. Temperature = 25° C.
2. Gases at one atmosphere pressure.
3. Dissolved substances at a concentration of one mole/liter. (Note that H^+ ions at a concentration of one mole/liter would make the *p*H of a system under standard conditions equal to zero.)
4. Pure liquids and solids in their most stable form at one atmosphere pressure and 25° C.

A change in free energy measured under these conditions is termed a "standard free energy change," symbolized by $\Delta G°$. Most of the values for change in free energy found in the chemical literature are $\Delta G°$ values.

Will the following reaction occur spontaneously under standard conditions?

_____yes
_____no

glucose-6-phosphate ⟶ fructose-6-phosphate

($\Delta G° = +400$ cal/mole)

1.56 ______

Change in free energy under cellular conditions Chemical reactions in a living cell generally occur under nonstandard conditions. For example, in a cell the *p*H is approximately 7, in contrast to the standard *p*H of 0. For this reason, the $\Delta G°$ measurement is not often used by biologists.

1.55
no (because $\Delta G°$ is positive)

It is possible to calculate change in free energy under nonstandard conditions from $\Delta G°$ by the following formula:

$$\Delta G = \Delta G^\circ + RT \ln \frac{[\text{products}]}{[\text{reactants}]}$$

where ΔG = change in free energy under nonstandard conditions
R = the gas constant
T = absolute temperature
[products] = concentration of products

(Do not try to memorize this formula; merely note that a change in free energy under nonstandard conditions can be calculated from ΔG° and vice versa.)

Biologists frequently work with a ΔG measured at a *p*H of 7, with temperature, concentration of species other than H^+, and pressure at their standard values. This ΔG is designated by the symbol $\Delta G'$. $\Delta G'$ is accurate enough for most purposes, so the convention will be adopted here and used throughout the remainder of this book.††

††You may find that some values for $\Delta G'$ found in this book do not agree exactly with those found in other books. This may be because the values have been rounded off in different ways or because some of the $\Delta G'$ values have been calculated for solutes at physiological concentrations, rather than at the standard concentration of one mole/liter.

1.57 ________________

Definitions of exergonic and endergonic Because "a reaction for which ΔG is positive (or negative)" is rather cumbersome to write, the more concise and convenient terms "exergonic" and "endergonic" are often used.

1.56
no answer

● Exergonic reaction: a reaction for which ΔG is negative.
Endergonic reaction: a reaction for which ΔG is positive.

Endergonic reactions
____occur spontaneously.
____do not occur spontaneously.

1.58 ________________

Which of the following four statements would a biochemist use to say that a certain reaction is exergonic under cellular conditions?
____ΔG° is positive.
____$\Delta G'$ is positive.

1.57
do not occur spontaneously

_____$\Delta G°$ is negative.
_____$\Delta G'$ is negative.

1.59 ____

Free energy diagrams Another convention that you will encounter later in this book is the free energy diagram for a reaction. One such diagram is shown here.

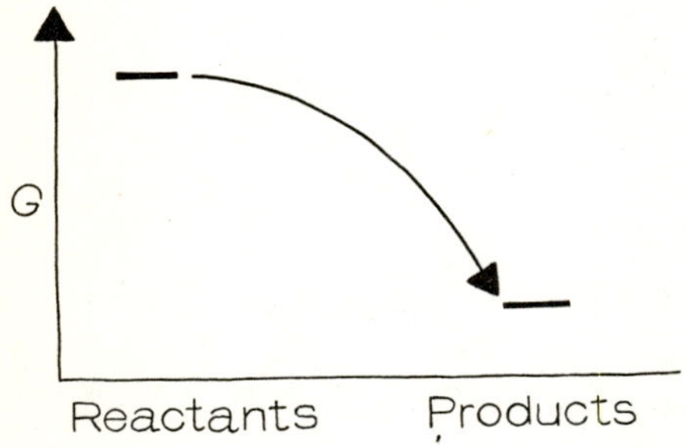

➡ If you already know how to construct diagrams of this type, skip to frame 1.60.

Explanation of the diagram

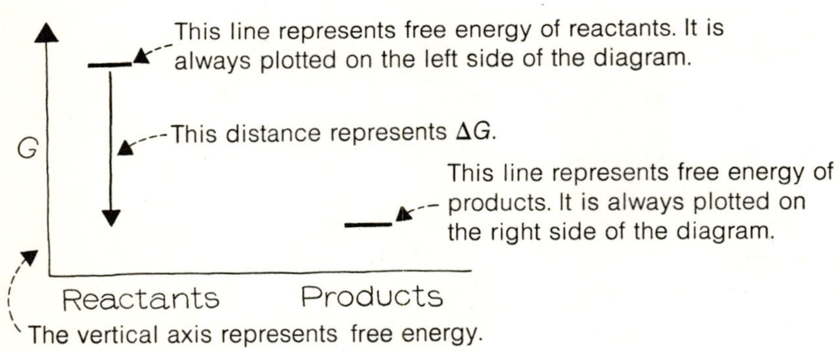

Does the first diagram represent an endergonic reaction or an exergonic reaction?

1.58
$\Delta G'$ is negative.

Teacher's Manual for

Guide to Cellular Energetics

by Lynne C. Carter

University of California, Berkeley

Contents

W. H. FREEMAN AND COMPANY
San Francisco

Printed in the United States of America.

Guide to Cellular Energetics is a supplement to lectures and course work in cell metabolism, particularly cellular energetics. It is intended for use in such courses as introductory biology, cell biology, cell physiology, molecular biology, and biochemistry. We assume that students who use this guide have taken or are currently taking general chemistry.

The aim of *Guide to Cellular Energetics* is to give students the basic chemical concepts needed to comprehend what glycolysis and respiration accomplish, by what strategy it is accomplished, and why it is necessary to the cell. A detailed statement of objectives is given on pages 2 - 5.

The guide deals with four major topics, which are presented in the context of cellular metabolism. These topics are free energy, how coupled reactions go forward, how reactions are speeded up, and oxidation/reduction reactions. *Guide to Cellular Energetics* fills a need for a compact and coherent presentation of these topics, which are presented diffusely in texts heretofore available. Although the strategy of glycolysis and respiration is a main theme of the guide, the actual reactions of glycolysis and respiration are presented with minimal discussion, in keeping with the supplementary nature of the guide and our expectation that teachers will want to present the pathways of glycolysis and respiration in the manner most appropriate for their particular students and teaching objectives.

The format of *Guide to Cellular Energetics* differs from that of most texts in two respects. As a reader progresses through the guide, he is presented with questions and problems and is asked to check his answers by comparing them to the answers presented elsewhere on the page. This active involvement with the subject matter allows the reader to see if he has mastered one fact or principle before going on to the next. In addition, the reader is given frequent opportunities to skip practice problems or material he already knows and also to branch to remedial material or supplementary material. The flexibility achieved through this approach makes the guide suitable for a diverse audience.

Guide to Cellular Energetics is designed as a self-teaching manual through which the student works independently. Instructions for working through the guide are included in the guide. The guide also contains a glossary, an index, a final test, and correct answers to the final test questions. A second, equivalent form of the test and an answer key are included in this manual, pages 5 - 14.

One to two weeks should be enough time for students who are taking a full course load to complete the guide. Students in an introductory biology course found that working through the guide required two to eight hours actual reading time, the average time being four hours.

During the development of *Guide to Cellular Energetics*, students from an introductory biology course and an introductory chemistry course read the manuscript and offered their evaluations and criticisms of the guide. Feedback from these students was used in revising early drafts of the guide.

OBJECTIVES OF *GUIDE TO CELLULAR ENERGETICS*

These objectives are stated as tasks that a student should be able to perform correctly after working through the guide. The objectives are arranged according to unit and topic.

Unit I. Predicting Spontaneous Change

A. Disorder, free energy, and spontaneous change

1. Given a description of a system before and after a change, predict whether that change is spontaneous. The types of system to be used are as follows.
   a. Simple isolated system. Change in positional disorder is indicated by a diagram.
   b. Isolated system composed of non-isolated system plus surroundings. Disorder of each component is assigned a numerical value.

2. Given a number of plausible alternatives, pick out the correct relationship between free energy and disorder.

B. Changes in disorder that result from cellular respiration and growth

1. From a number of alternatives, choose the correct explanation of how respiration makes it possible for an organism to grow; <u>or</u> choose the diagram that correctly expresses changes in disorder that result from simultaneous respiration and growth.

C. Interpretation of ΔG values

1. Given a reaction and its ΔG, predict whether or not that reaction will occur spontaneously.

D. Symbols, definitions, and conventions pertaining to free energy

1. For each of the following symbols, supply the definition; <u>or</u> given sets of possible reaction conditions, match each symbol to appropriate set(s) of conditions: ΔG, ΔG°, and $\Delta G'$.

2. Given ΔG for a reaction, construct a conventional free energy diagram for the reaction.

3. Given ΔG for a reaction, predict ΔG for the reverse reaction.

E. Factors affecting the value of ΔG

1. Given two reactions which differ with respect to one of the following factors,
   a. temperature
   b. concentration of reactants or products
   c. rate at which the reaction proceeds

d. pathway by which the reaction proceeds

predict whether or not the two reactions have the same value of ΔG.

Unit II. Coupled Reactions

A. Mechanism of coupling

1. Given two reactions which can be coupled and a number of possible explanations for why the endergonic reaction goes forward, select the correct explanation.

B. Requirements for coupling

1. Given two or more reactions and their ΔG's, predict whether or not they can be coupled. The following should be included in the question.
   a. Two reactions that can be coupled
   b. Three reactions that can be coupled
   c. Two reactions with a common intermediate and a net positive ΔG
   d. Two reactions with a net negative ΔG and a common starting material, but no common intermediate

C. Notation and terminology

1. Given examples and non-examples of coupled reactions, identify the coupled reactions.

2. Rewrite, using conventional curved arrows notation, two separate reactions that can be coupled.

3. Given the information that a reaction is exergonic or endergonic, predict whether or not it will occur spontaneously.

Unit III. Rate of Reaction

A. Factors affecting rate of reaction

1. Given a reaction, its ΔG, and directions to indicate the relative rate of the reaction, recognize that the correct answer choice is, "insufficient information available to determine the rate of the reaction."

2. Given a number of possible relationships, recognize the correct relationship between <u>spontaneous</u> and <u>fast</u>.

3. List the events that must take place between two reactant molecules before they can become products.

4. From an array of definitions, pick out the correct definition of free energy of activation; <u>or</u> on a free energy diagram for a reaction, select the distance which represents free energy of activation.

B. Speeding up reactions

1. Given a number of possible explanations, recognize the correct explanation for why rise in temperature speeds up a reaction or why drop in temperature slows down a reaction.

2. From a list of possible properties of catalysts, select those which actually are properties of catalysts. The list of possible properties should include the following:
   a. effect on overall ΔG of the reaction
   b. effect on free energy of activation
   c. whether catalyst is used up or altered permanently by the reaction

3. Given a number of possible explanations, choose the correct explanation for how a catalyst speeds up reactions; <u>or</u> given written descriptions of interactions, discriminate between examples and non-examples of catalysis.

C. Enzymes

1. Given a number of possible relationships (represented by diagrams or written descriptions), recognize the correct relationship between enzymes and catalysts.

2. From a list of possible properties of enzymes, select those which actually are properties of enzymes. The list should include the following:
   a. specificity
   b. composition
   c. structure
   d. interaction with substrate

3. Given a number of possible alternatives, recognize the advantages of enzymes over ordinary catalysts; <u>or</u> recognize the methods by which enzymes effect catalysis.

Unit IV. Oxidation/Reduction Reactions

A. Definitions

1. Given a collection of chemical equations, discriminate between redox reactions and non-redox reactions. The collection of equations should include the following:
   a. redox reactions in which hydrogen atoms are transferred
   b. non-redox reactions involving ions

2. Given a half reaction, state whether it is an oxidation or a reduction.

3. Given a redox reaction, identify the oxidizing agent or the reducing agent.

B. Notation

1. Compounds

   a. When writing equations, employ from memory the conventional notation for the reduced form of NAD^+.

   b. Identify forms of NAD^+, FAD, and cytochromes as oxidized or reduced.

2. Reactions

   a. Given a redox reaction written in the conventional curved arrows notation, rewrite it as two half reactions.

   b. Given a redox reaction written as a chemical equation or written in curved arrows notation, describe the direction in which electrons are transferred.

C. Redox potential

1. Given the relative electron affinities of two compounds, predict the direction in which electrons will flow spontaneously; <u>or</u> given two compounds and the direction in which electrons flow spontaneously, deduce the relative electron affinities of the compounds.

2. Answer questions requiring memory of the relation between relative value of E_0 and electron affinity, for example, "Which accepts electrons more readily, compound A (has positive E_0) or compound B (has negative E_0)?"

D. Redox reactions in respiration

1. Discriminate between correct incorrect statements concerning redox reactions in respiration. Statements covering the following topics should be included:
   a. direction of electron flow
   b. components of the electron transport chain
   c. whether redox reactions in the electron transport chain are endergonic or exergonic.

STUDENT TEST

The student test, which begins on the following page, is equivalent to the final test included in <u>Guide to Cellular Energetics</u>. Both tests are based on the preceding objectives of the guide. The answer key, found on page 14 of this manual, gives the correct answer to each question and cites the unit and section of the guide on which each question is based.

_______ 1. Will the following change occur spontaneously? (Assume that this is an isolated system and that kinetic and potential energy are not altered during the change.)

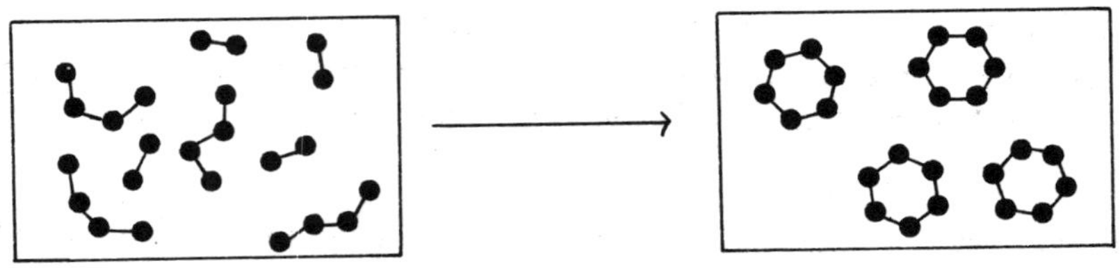

_______ 2. Will the following change occur spontaneously?

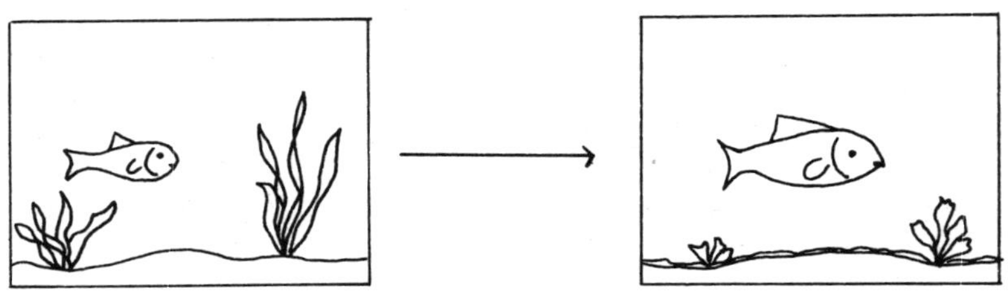

Before change: disorder of fish (a nonisolated system) = 200; disorder of surroundings = 250.

After change: disorder of fish = 150; disorder of surroundings = 350.

3. Briefly explain what each of the following symbols stands for.

ΔG

$\Delta G'$

ΔG°

4. If a reaction occurs spontaneously, it is

___endergonic.
___exergonic.

______ 5. Which of the following statements correctly expresses the relationship between disorder and free energy?

a. Free energy of a nonisolated system decreases as the disorder of the nonisolated system decreases.

b. Free energy of a nonisolated system decreases as the disorder of the nonisolated system increases.

c. Free energy of a nonisolated system decreases as the disorder of (system + surroundings) decreases.

d. Free energy of a nonisolated system decreases as the disorder of (system + surroundings) increases.

______ 6. Of the following three diagrams, which is the closest analogy to how glycolysis and respiration make growth possible?

a.

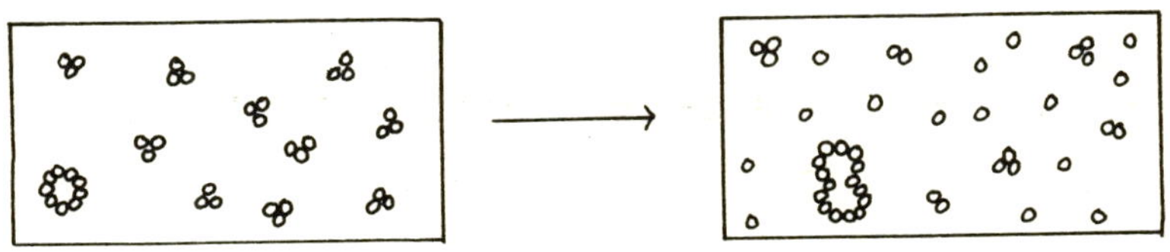

b.

c.

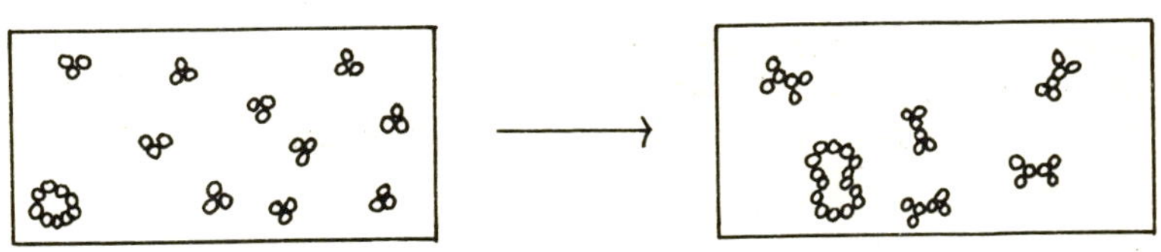

______ 7. Will the following reaction occur spontaneously?

acetyl phosphate + $H_2O \longrightarrow$ acetic acid + H_3PO_4

(ΔG = -10,000 cal/mole)

8. Draw a free energy diagram for the reaction in question 7.

9. $\Delta G°$ for the following reaction is -5,000 cal/mole.

$$\text{sucrose} + H_2O \longrightarrow \text{glucose} + \text{fructose}$$

For the reaction, glucose + fructose $\longrightarrow$ sucrose + H_2O, the value of $\Delta G°$

___cannot be predicted from the information given.
___is ______________ cal/mole.

10. Indicate whether the $\Delta G'$s of the following pairs of reactions will be the same or different. Temperatures appear at right.

_______ a.

oxalacetic acid (.3 molar) $\xrightarrow{\text{2 H atoms}}$ malic acid (.2 molar) — 29° C

oxalacetic acid (.3 molar) $\xrightarrow{CO_2}$ pyruvic acid (.1 molar) $\xrightarrow{CO_2 \text{ + 2 H atoms}}$ malic acid (.2 molar) — 29°C

_______ b.

acetic acid (2 molar) + H_3PO_4 (1 molar) $\longrightarrow$ acetyl phosphate (2 molar) + H_2O — 20°C

acetic acid (2 molar) + H_3PO_4 (1 molar) $\longrightarrow$ acetyl phosphate (2 molar) + H_2O — 10°C

_______ c.

ethanol (.02 molar) $\longrightarrow$ CO_2 (1 atm.) + acetaldehyde (.04 molar) — 40°C

ethanol (.03 molar) $\longrightarrow$ CO_2 (1 atm.) + acetaldehyde (.04 molar) — 40° C

_______ d.

succinic acid (.2 molar) + FAD (.1 molar) $\xrightarrow{\text{in the presence of an enzyme that speeds up the reaction}}$ fumaric acid (.1 molar) + $FADH_2$ (.1 molar) — 20°C

succinic acid (.2 molar) + FAD (.1 molar) $\xrightarrow{\text{no enzyme present}}$ fumaric acid (.1 molar) + $FADH_2$ (.1 molar) — 20°C

11. Rewrite the following pair of coupled reactions in the conventional notation used for coupled reactions.

$$\text{ATP} + H_2O \longrightarrow \text{ADP} + H_3PO_4$$

$$H_3PO_4 + \text{glycerol} \longrightarrow \text{glycerol-1-phosphate} + H_2O$$

______ 12. Following are four sets of reactions. In which set(s) can the reactions shown be coupled to each other? ($\Delta G'$ in calories/mole appears at right.)

a. acetic acid + glyoxylic acid ⟶ malic acid　　-2,000
malic acid ⟶ oxalacetic acid + 2 H　　+6,700

b. oxalacetic acid + 2 H ⟶ malic acid　　-6,700
malic acid ⟶ fumaric acid + H_2O　　+750
fumaric acid + 2 H ⟶ succinic acid　　-4,000

c. glucose-6-phosphate ⟶ fructose-6-phosphate　　+400
glucose-6-phosphate + H_2O ⟶ glucose + H_3PO_4　　-3,300

d. acetyl phosphate + H_2O ⟶ acetic acid + H_3PO_4　　-10,000
acetic acid + 2 H ⟶ acetaldehyde + H_2O　　+8,500

______ 13. Which of the following represent(s) coupled reactions?

a. glucose ⟶ glucose-6-phosphate
ATP ⟶ ADP

b. fructose-1,6-diphosphate ⟶ dihydroxyacetone phosphate + glyceraldehyde-3-phosphate

c. fumaric acid $\xrightarrow{H_2O}$ malic acid $\xrightarrow{2\ H}$ oxalacetic acid

d. citric acid ⇌ isocitric acid

______ 14. Following the pair of coupled reactions that appear below is a list of possible reasons to explain why reaction 1 proceeds. Which explanation is most correct?

1. malic acid ⟶ oxalacetic acid + 2 H　$\Delta G' = +6{,}700$ cal/mole
2. oxalacetic acid ⟶ pyruvic acid + CO_2　$\Delta G' = -7{,}000$ cal/mole

Possible explanations:

a. Reaction 1 proceeds because it is endergonic.
b. Reaction 1 proceeds because reaction 2 spontaneously consumes a product of reaction 1.
c. Reaction 1 proceeds because it floods reaction 2 with oxalacetic acid, which is a reactant of 2.
d. Reaction 1 proceeds because it uses up malic acid.

15. The following reaction occurs
___ at a relatively fast rate.
___ at a relatively slow rate.
___ Not enough information is given.

fructose-6-phosphate ⟶ glucose-6-phosphate　$\Delta G' = -400$ cal/mole

______ 16. Which statement correctly expresses the relationship between <u>spontaneous</u> and <u>fast</u> reactions?

a. A spontaneous reaction is the same thing as a fast reaction.
b. Some, but not all, spontaneous reactions are fast.
c. Spontaneous reaction and fast reaction are mutually exclusive terms.
d. A spontaneous reaction is the opposite of a fast reaction.

17. In any reaction mixture, only a fraction of the encounters between reactant molecules result in products being formed. Most of the encounters are unsuccessful. Briefly describe what factor or factors might account for failure to form products in these unsuccessful encounters.

______ 18. Which is the most correct definition of free energy of activation?

a. The difference between the free energy of reactants and free energy of products.
b. The increase in energy of the molecules when the temperature of a reaction mixture is raised 10 degrees C.
c. The difference between the free energy of reactants and the free energy of the transition state.
d. The difference between the free energy of products and the free energy of the transition state.
e. The free energy necessary to promote an electron to a higher energy level.

______ 19. If we lower the temperature of a reaction mixture, the reaction goes slower because (choose one or more)

a. the free energy of activation has been reduced.
b. the rate of collisions has decreased.
c. the molecules are moving slower and the system is less disordered.
d. the kinetic energy of the molecules is lowered enough that converting kinetic energy to potential energy decreases the disorder of the system.

______ 20. For the hypothetical reaction, 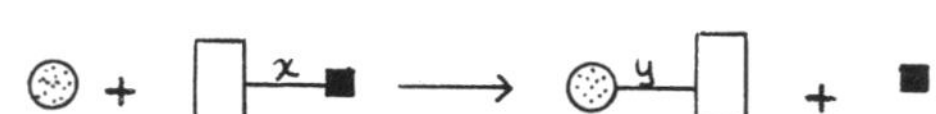

which of the following interactions represent(s) catalysis?

a. Strengthening bond x.
b. Weakening bond x.
c. Strengthening bond y.
d. Weakening bond y.
e. Making  less reactive toward 
f. Making  more reactive toward 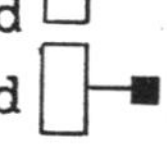

_______ 21. Which statement(s) regarding catalysts is (are) true?

a. Catalysts raise the ΔG of a reaction.
b. Catalysts are not altered permanently or used up by the reactions they catalyze.
c. Catalysts decrease the kinetic energy of the reactant molecules.
d. Catalysts raise the free energy of activation of a reaction.

_______ 22. Which diagram most correctly expresses the relationship between enzymes and catalysts?

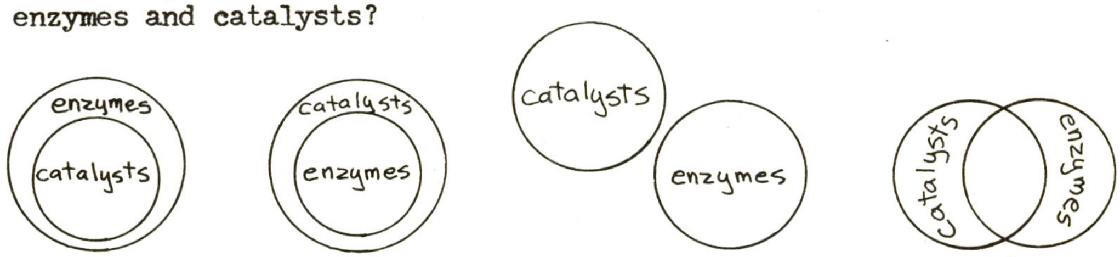

a. b. c. d.

_______ 23. Which statement(s) about enzymes is (are) true?

a. One enzyme usually catalyzes only one or a few reactions.
b. Enzymes are composed mostly of protein.
c. The three-dimensional structure of an enzyme is a straight thread.
d. The three-dimensional structure of an enzyme is a thread coiled into a double helix.
e. Chemical bonds are formed between an enzyme and its substrate.

_______ 24. The decomposition of hypothetical compound X may be speeded up by adding H^+ and S^{-2} (catalysts) or by adding an enzyme. The reaction goes much faster in the presence of enzyme than it does in the presence of H^+ and S^{-2}. Which of the following explain(s) why the enzyme can make the reaction go faster than H^+ and S^{-2} can?

a. The enzyme can operate at a lower temperature.
b. In the enzyme, catalytic groups are held in optimum orientation relative to the substrate.
c. The enzyme can supply more energy to the molecule of X than H^+ or S^{-2} can.
d. None of the above.

_______ 25. Which of the following are redox reactions? (Choose one or more.)

a. $Zn^{++}SO_4^{--} + Fe \longrightarrow Fe^{++}SO_4^{--} + Zn$
b. sucrose + H_2O $\longrightarrow$ glucose + fructose
c. malic acid $\longrightarrow$ oxalacetic acid + 2 H atoms
d. $Na^+OH^- + H^+Cl^- \longrightarrow Na^+Cl^- + H^+OH^-$

26. The following reaction is
___an oxidation.
___a reduction.

$$Cu^{++} + e^{-} \longrightarrow Cu^{+}$$

27. Indicate whether the form given for each of the following compounds is the oxidized or the reduced form of that compound.

a. $FADH_2$
___oxidized
___reduced

b. cytochrome c-Fe^{+++}
___oxidized
___reduced

c. NAD^{+}
___oxidized
___reduced

________ 28. Identify the reducing agent in the following reaction.

$$\frac{1}{2}\,O_2 + 2\,H^{+} + 2\text{ cytochrome a-}Fe^{++} \longrightarrow H_2O + 2\text{ cytochrome a-}Fe^{+++}$$

29. Complete the following equation:

$$NAD^{+} + FADH_2 \longrightarrow \underline{\hspace{6cm}}$$

30. Rewrite the following as two half reactions:

31. In the following reaction, electrons are transferred from
___cytochrome b-Fe^{++}
___cytochrome b-Fe^{+++}
___riboflavin + 2 H^{+}
___riboflavin-H_2
and ultimately become part of
___cytochrome b-Fe^{++}
___cytochrome b-Fe^{+++}
___riboflavin + 2 H^{+}
___riboflavin-H_2

$$\text{cytochrome b-}Fe^{++} + \text{riboflavin} + 2\,H^{+} \longrightarrow \text{cytochrome b-}Fe^{+++} + \text{riboflavin-}H_2$$

32. The flow of electrons pictured below

___will take place spontaneously.
___will not take place spontaneously.

compound with high electron affinity $\xrightarrow{\text{path of electrons}}$ compound with low electron affinity

33. The more positive the E_0 of a compound,
___the greater its electron affinity.
___the lower its electron affinity.

_______ 34. Which of the following statements is (are) true?

a. The redox reactions in the electron transport chain are exergonic.
b. The last substance to accept electrons in the electron transport chain is oxygen.
c. During electron transport, electrons are transferred to ADP, converting it to ATP.
d. None of the above.

KEY

Unit and section of guide on which question based	Question number	Correct answer
1,1	1.	no
1,2	2.	yes
1,3	3.	ΔG = change in free energy; free energy of products minus free energy of reactants $\Delta G'$ = change in free energy at pH 7, other conditions standard ΔG° = change in free energy under standard conditions
1,3	4.	exergonic
1,2	5.	d
1,2	6.	a
1,3	7.	yes
1,3	8.	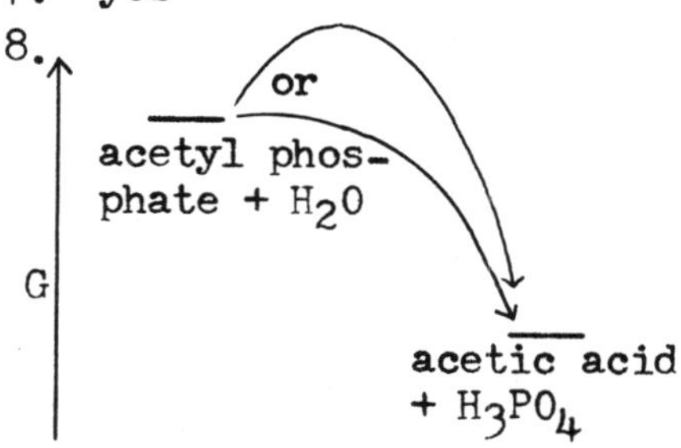
1,3	9.	is +5,000 cal/mole
1,3	10.	a. same b. different c. different d. same
2,3	11.	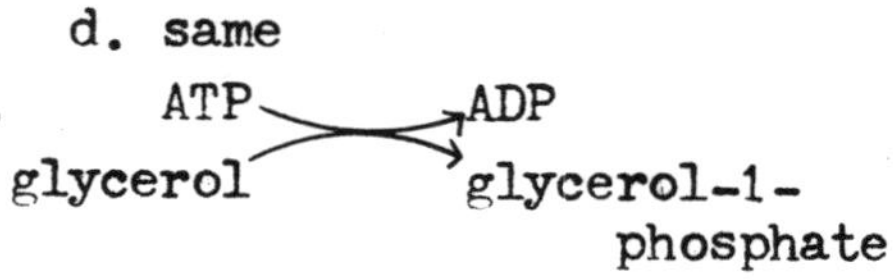
2,2	12.	b,d
2,3	13.	a,c
2,1	14.	b
3,1	15.	Not enough information is given.

Unit and section of guide on which question based	Question number	Correct answer
Overview	16.	b
3,2	17.	Reactants did not collide in proper orientation; reactants did not have enough energy for disrupting old bonds and forming new bonds (i.e. not enough energy to form the transition state).
3,2	18.	c
3,3	19.	b,d
3,3	20.	b,f
3,3	21.	b
3,4	22.	b
3,4	23.	a,b,e
3,4	24.	b
4,3	25.	a,c
4,2	26.	a reduction
4,3	27.	a. reduced b. oxidized c. oxidized
4,3	28.	2 cytochrome a-Fe^{++}
4,3	29.	NADH + H^+ +FAD
4,3	30.	pyruvic acid + 2 H^+ + 2 e^- $\longrightarrow$ lactic acid $FADH_2 \longrightarrow$ FAD + 2 H^+ + 2 e^-
4,3	31.	cytochrome-b-Fe^{++} riboflavin-H_2
4,5	32.	will not take place spontaneously
4,5	33.	the greater its electron affinity
4,6	34.	a,b

1.60

Draw a free energy diagram representing a reaction for which ΔG is positive.

1.59
exergonic reaction

1.61

a. The diagram in frame 1.60 represents an
____exergonic reaction.
____endergonic reaction.

b. The reaction diagramed in frame 1.60
____will occur spontaneously.
____will not occur spontaneously.

1.60

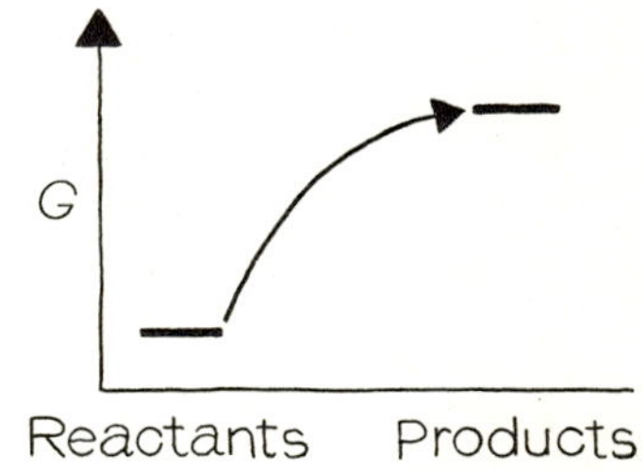

summary

1.62

You have now completed Section Three. The content of this section is summarized in the following questions.

1.61
a. endergonic reaction
b. will not occur spontaneously

a. Free energy is symbolized by ____.

b. Change in free energy is symbolized by ____, which equals $G_{\text{products}} - G_{\text{reactants}}$.

c. A reaction occurs spontaneously if ΔG for that reaction is
____positive.
____negative.

d. ΔG for the reverse of a reaction equals ΔG of the forward reaction multiplied by ____.

e. What two variables affect the value of ΔG?

f. What two variables do not affect the value of ΔG?

g. In what units is ΔG measured?

h. ____ stands for change in free energy under standard conditions.

i. ____ stands for change in free energy at standard temperature, pressure, and concentration, but at $pH = 7$.

j. An exergonic reaction is a reaction for which ΔG is
____positive.
____negative.

k. An endergonic reaction is a reaction for which ΔG is
____positive.
____negative.

l. It is standard practice to diagram the change in free energy for a reaction by plotting $G_{\text{reactants}}$ and G_{products} on a vertical scale of free energy. $G_{\text{reactants}}$ is always plotted to the
____left of G_{products}.
____right of G_{products}.

You have now completed Unit One. Go to Unit Two.

help frame for unit one

1.63

1.08

As soon as you think you can answer the question in frame 1.08, jump back to that frame. It is better to avoid going through this entire help frame if you do not need to.

a. The probability that a given situation will occur equals:

$$\frac{\text{Number of arrangements that correspond to the situation}}{\_\_\_\_\_\_\_\_\_\_}$$

(fill in the denominator)

1.62

a. G
b. ΔG
c. negative
d. -1
e. the temperature at which the reaction is carried out and the concentrations of products and reactants

b. The number of arrangements that correspond to an orderly situation will be
____many.
____few.
Therefore, the probability that an orderly situation will occur is a small number divided by the total.

c. The probability that a disordered situation will occur is
____a large number divided by the total.
____a small number divided by the total.

● Go back to frame 1.08 and try to answer the question asked there.

f. the rate at which the reaction takes place and the pathway by which the reaction occurs
g. calories/mole (or kilocalories/mole)
h. $\Delta G°$
i. $\Delta G'$
j. negative
k. positive
l. left of G_{products}

1.63

a. total number of possible arrangements
b. few
c. a large number divided by the total

unit two

coupled reactions: how endergonic reactions proceed

one

conditions under which endergonic reactions proceed

2.01

In "An Overview of Cellular Metabolism" you read that many reactions that are vital to the life of an organism do not proceed spontaneously; in other words, they are endergonic. For example, the synthesis of proteins from amino acids is an endergonic reaction, as are syntheses of other important macromolecules.

However, a reaction with a positive $\Delta G'$ cannot achieve synthesis of products independently. Only reactions in which there is a decrease in free energy occur spontaneously. Therefore, the cell as a system must somehow undergo a net decrease in free energy as these endergonic reactions proceed.

What kind(s) of system could undergo an overall decrease in free energy at the same time that a reaction with a positive $\Delta G'$ is occurring? Two such systems will be described in the frames that follow. From these two examples we shall gain an understanding of how, in the cell, reactions with positive $\Delta G'$s can proceed in the direction of product formation.

2.02

Consider a system made up of the following two reactions:

$$A \longrightarrow B \qquad (\Delta G' = +1{,}060 \text{ cal/mole})$$

$$B \longrightarrow C \qquad (\Delta G' = -1{,}150 \text{ cal/mole})$$

A microscopic view of this system Since $\Delta G'$ for $A \longrightarrow B$ is positive, the reverse reaction ($B \longrightarrow A$) occurs spontaneously. In a mixture in which A and B are at equilibrium, the amount of A greatly exceeds the amount of B.

We can represent the equilibrium situation with the following model:

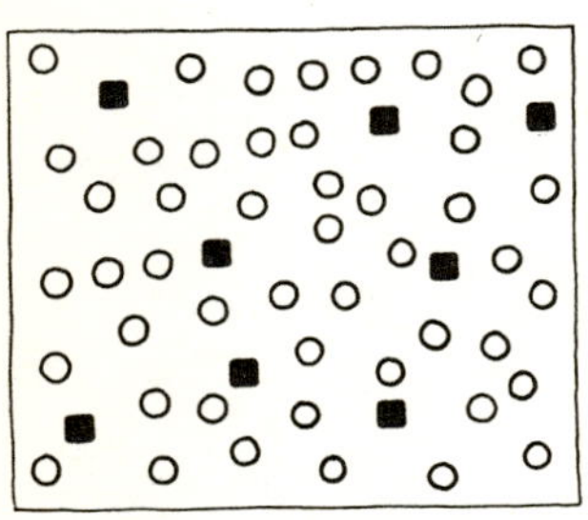

○ = *A* ■ = *B*

Note that there are six molecules of A for every molecule of B. This ratio of A to B will prevail whenever the system is in equilibrium.

What will happen if B is somehow removed from the system?

● If you need help in answering this question, go to frame 2.39.

$\Delta G'$ for $B \longrightarrow C$ (the second reaction in the system) is negative, and at equilibrium the ratio of B to C is 1 to 7. If we have a system where the first reaction has come to equilibrium

2.01
no answer

and if we then let the second reaction take place, *B* is spontaneously converted to *C*, as shown in the following diagram:

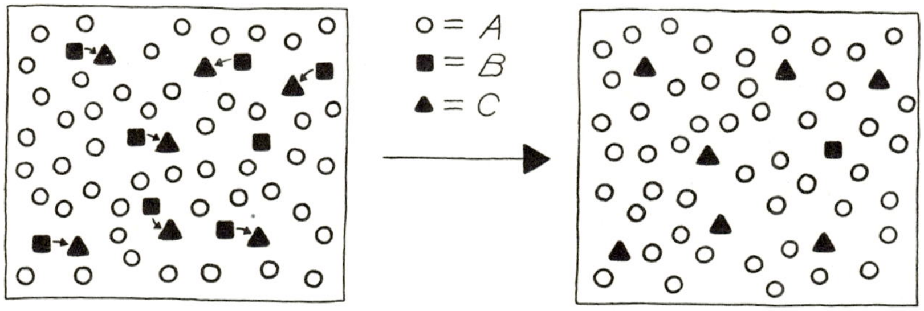

Ratio of *A* to *B* is now 48 to 1
Ratio of *B* to C is 1 to 7

Since *B* is drained away by the second reaction, the balance between *B* and *A* has been upset. To compensate and bring the ratio of *A* to *B* back to 6 to 1, some *A* will be converted to *B*. This is shown in the following diagram:

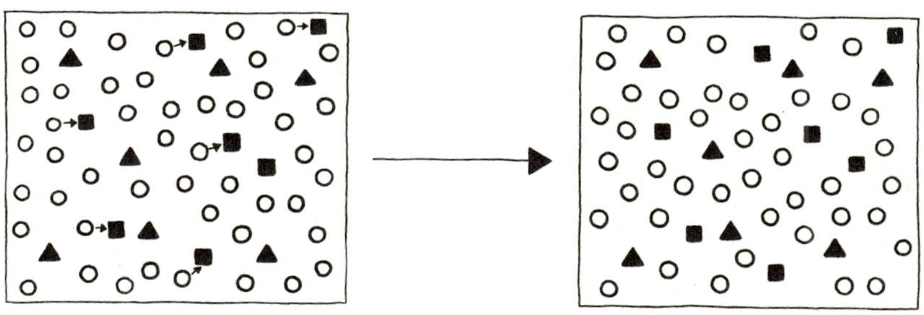

Ratio of *A* to *B* is now 6 to 1
Ratio of *B* to C is now 1 to 1

The ratio of *B* to *C* is thus changed so that *B* will again be drawn off by the second reaction and converted to *C*. *A* will consequently be converted to *B*, and so on. The net result is that much *A* is converted to *B*; in other words, the reaction $A \longrightarrow B$ occurs. This would not have happened if the second reaction were not part of the system.

2.03

A macroscopic view of the same system The reason why the endergonic reaction $A \longrightarrow B$ proceeds can also be explained in the following terms. The decrease in free energy of the

2.02
Some *A* will be converted to

system as B goes to C outweighs the increase in free energy as A goes to B. As you can see in the next diagram, the net $\Delta G'$ of the system is negative, and both reactions occur.

B so that the ratio of A to B is again 6 to 1.

Go to frame 2.39 if you did not get the correct answer.

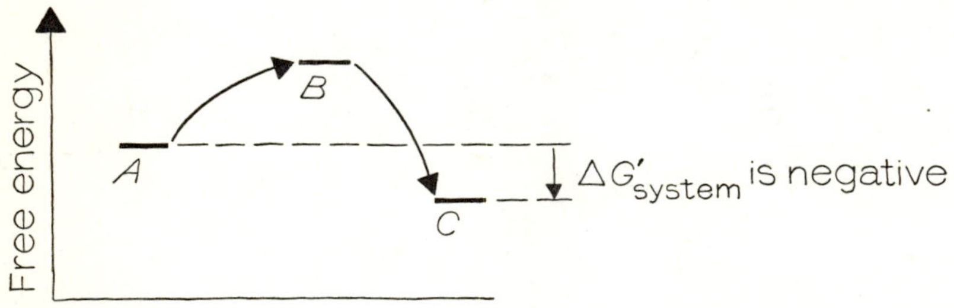

Note the necessary properties of the exergonic reaction in this system:

1. It uses up a product of the reaction with the positive $\Delta G'$.
2. Its $\Delta G'$ is negative enough that $\Delta G'_{system}$ (which equals the sum of the $\Delta G'$s of the two reactions) is negative.

2.04

We have just considered one kind of system that will undergo an overall decrease in free energy at the same time that a reaction with a positive $\Delta G'$ is occurring. Let us look briefly at a second kind of system that undergoes an overall decrease in free energy while including a reaction with a positive $\Delta G'$.

2.03
no answer

A microscopic view of the second system Again a model represents the equilibrium situation for a reaction with a positive $\Delta G'$:

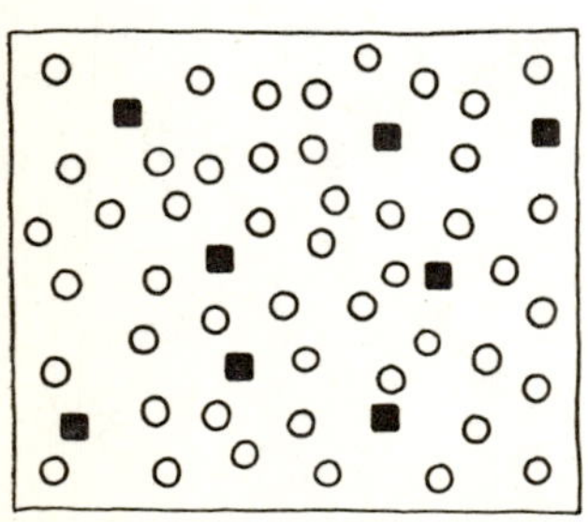

O = A ■ = B

$A \longrightarrow B$ $\Delta G' = +1{,}060$ cal/mole
Ratio of A to B at equilibrium is 6 to 1

In the system we considered in frame 2.02, A was converted spontaneously to B because the equilibrium was upset when B was drained away by the spontaneous reaction $B \longrightarrow C$.

Can you think of another way to upset the equilibrium between A and B, causing A to be converted to B?

● If so, write your answer in the space provided and check it in the usual way. If not, go to frame 2.05.

2.05

Imagine that the system contains, in addition to the reaction $A \longrightarrow B$, the reaction $X \longrightarrow A$, which has a negative $\Delta G'$. This would have the effect of flooding the $A \longrightarrow B$ reaction with A:

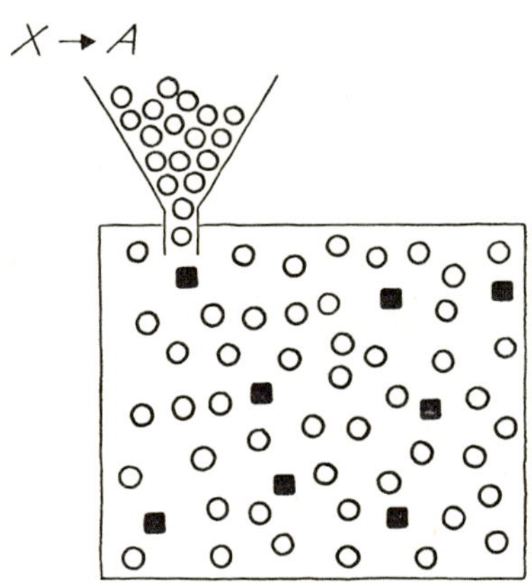

Will this reaction unbalance the equilibrium between A and B?

_____yes

_____no

2.04

add more A to the box (for example, by a spontaneous reaction $X \longrightarrow A$)

● If you did not get this answer or its equivalent, go to frame 2.05. If you got this answer or its equivalent, go to frame 2.07.

2.06

What will happen to restore the balance between A and B?

2.05

yes

2.07
2.04

The preceding three frames can be summarized as follows:

An endergonic reaction can proceed if it is part of a system that also contains a reaction that (1) produces a reactant of the endergonic reaction; and (2) has a $\Delta G'$ negative enough that $\Delta G'_{\text{system}}$ is negative.

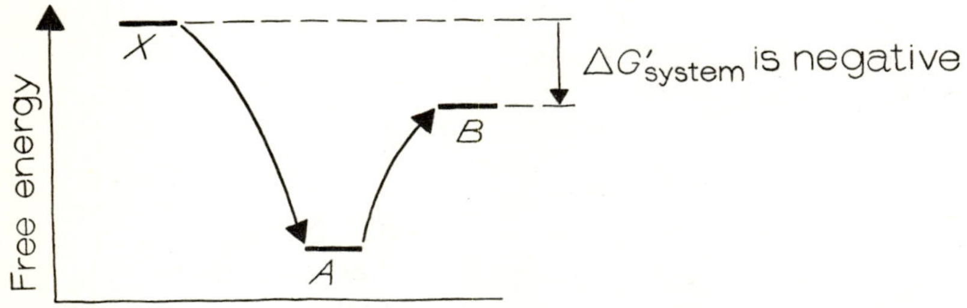

2.06
A will be converted to *B*.

summary

2.08

An endergonic reaction does not proceed spontaneously if it is the only reaction in an isolated system. However, if the system is enlarged to include an additional reaction with certain properties, the endergonic reaction will be driven forward.

What properties must the additional reaction possess to drive forward the endergonic reaction?

2.07
no answer

● Refer back to frames 2.03 and 2.07 if you need help in answering this question.

2.08
The additional reaction must consume a product or produce a reactant of the endergonic reaction, and the additional reaction must have a ΔG negative enough that ΔG of the system is negative.

two

terminology

Now that you have been introduced to the fundamental ideas of this unit, it is appropriate to introduce you to some conventional terminology.

2.09 ______

Coupled reactions

- When an endergonic reaction is driven forward by an exergonic reaction, the two reactions are said to be coupled.

➡ If you already know what a common intermediate is, you may skip to frame 2.13. If you cannot answer the questions there, return to frame 2.10.

2.10 ______

Common intermediate

- If a product of one reaction is a reactant in another reaction, the two reactions share a common intermediate

2.09
no answer

B is the common intermediate of the following two reactions:

$$A \longrightarrow B$$
$$B \longrightarrow C$$

2.11 ______________________________

Do the following reactions share a common intermediate? If so, what is it?

____

$$A + B \longrightarrow C + D$$
$$W + C \longrightarrow X + Y$$

2.10
no answer

2.12 ______________________________

What is the common intermediate of the following two reactions?* (Note that you do not need to know the structures of these compounds to answer this question.)

____

$$ATP + H_2O \longrightarrow ADP + H_3PO_4$$
$$glucose + H_3PO_4 \longrightarrow glucose\text{-1-phosphate} + H_2O$$

*In these reactions water fits the definition of a common intermediate. However, the cell is essentially an aqueous system, and water molecules are very plentiful. Adding or drawing off molecules of water will not drive any reactions forward appreciably. Therefore, water is not usually considered to be a common intermediate in cellular reactions.

2.11
yes, *C*

2.13 ______________________________
2.09

Do the following two reactions have a common intermediate?
_____yes
_____no

$$L + M \longrightarrow P + Q$$
$$X + Y \longrightarrow M$$

2.12
H_3PO_4

2.14 ______________________________

The two reactions in frame 2.13 do have a common intermediate, *M*:

$$L + \textcircled{M} \longrightarrow P + Q$$
$$X + Y \longrightarrow \textcircled{M}$$

These reactions have been written in an order different from

2.13
yes

➡ If you answered correctly, go to frame 2.15. If you did not, go to frame 2.14.

that used earlier in this book, but this order does not alter the fact that a product of one reaction is a reactant in the other reaction.

2.15 ______

2.13

Two reactions have a common intermediate if ______
______.

2.14
no answer

2.16 ______

Examine the following pairs of reactions. For each pair, state whether the two reactions have a common intermediate, and if so, circle it.

a.

$$\text{glucose} + \text{fructose} \longrightarrow \text{sucrose} + H_2O$$
$$\text{glucose} + \text{galactose} \longrightarrow \text{lactose} + H_2O$$

___

b.

$$\text{pyruvic acid} \longrightarrow \text{acetaldehyde} + CO_2$$
$$2\text{ "H"} + \text{acetaldehyde} \longrightarrow \text{ethanol}$$

___

c.

$$\text{fumaric acid} + H_2O \longrightarrow \text{malic acid}$$
$$\text{pyruvic acid} + CO_2 + 2\text{ "H"} \longrightarrow \text{malic acid}$$

___

d.

$$\text{fumaric acid} + H_2O \longrightarrow \text{malic acid}$$
$$\text{succinic acid} \longrightarrow \text{fumaric acid} + 2\text{ "H"}$$

___

2.15
a product of one reaction is a reactant in the other reaction

2.17 ______

Practice in recognizing the conditions for coupling The basic conclusions of Section One can be restated, using the new terms you have just learned.

- Reactions can be coupled if
  1. they have a common intermediate, and
  2. the sum of their ΔGs is negative.

➡ If you have previously studied coupled reactions and have had practice in recognizing whether two reactions can be coupled, you may skip to frame 2.21.

2.16
a. no
b. yes, acetaldehyde
c. no
d. yes, fumaric acid

2.18 ______

Could the following reactions be coupled? Why or why not?
______

glycerol-1-phosphate + H_2O ⟶ glycerol + H_3PO_4

$(\Delta G' = -2{,}300 \text{ cal/mole})$

fructose + H_3PO_4 ⟶ fructose-6-phosphate + H_2O

$(\Delta G' = +3{,}800 \text{ cal/mole})$

2.17
no answer

2.19 ______

Which reaction (2, 3, or 4) could drive forward reaction 1?
______

1. lactic acid ⟶ pyruvic acid + 2 "H"
   $(\Delta G' = +6{,}000 \text{ cal/mole})$
2. pyruvic acid ⟶ acetaldehyde + CO_2
   $(\Delta G' = -4{,}700 \text{ cal/mole})$
3. phosphoenolpyruvic acid + H_2O ⟶ pyruvic acid + H_3PO_4
   $(\Delta G' = -12{,}800 \text{ cal/mole})$
4. pyruvic acid + H_2O ⟶ H_2 + CO_2 + acetic acid
   $(\Delta G' = -13{,}000 \text{ cal/mole})$

2.18
no, because the sum of ΔG's is not negative

2.20 ______

State why each of the other reactions in the preceding frame is not suitable for driving forward reaction 1.

2.19
4

2.21 ______
2.17

More than two reactions can be coupled It is important to realize that the number of reactions that can be coupled is not limited to two. In the cell there frequently occur long chains of reactions where the product of one reaction is used up as a reactant in the next step. As long as the sum of the ΔGs is negative, all the reactions will proceed.

2.20
The sum of the ΔG's for reactions 1 and 2 is not negative. Reaction 3 does not have a common intermediate with reaction 1.

2.22 ______

Consider the following series of three coupled reactions:

1. $A + B \longrightarrow C$ $(\Delta G' = +3{,}000 \text{ cal/mole})$
2. $C \longrightarrow D + E$ $(\Delta G' = -15{,}000 \text{ cal/mole})$
3. $D + F \longrightarrow G + H$ $(\Delta G' = +400 \text{ cal/mole})$

Which of the following statements is true?

____

1. Reaction 1 proceeds because reaction 2 consumes *C*.
2. Reaction 2 proceeds because reaction 1 floods the system with large quantities of *C*.
3. Reaction 2 proceeds because reaction 3 drains off *D*.

2.21
no answer

2.23 ______

a. Why does reaction 2 in frame 2.22 proceed?

b. Why does reaction 3 proceed?

2.22
1

summary

2.24 ______

● Two or more reactions can be coupled if

1. they have a common intermediate, and
2. the sum of their ΔGs is negative.

Two reactions share a common intermediate if the product of one reaction is a reactant in the other reaction.

2.23
a. because it has a negative $\Delta G'$, because it is exergonic, and so on
b. because reaction 2 floods it with *D* (or an answer using words to that effect)

2.24
no answer

three

notation for coupled reactions

Because coupled reactions are a common occurrence in the chemistry of the cell, a shorthand notation has been developed for writing them. This section will introduce the notation to you.

2.25 ____

Notation for a pair of coupled reactions What is the common intermediate of the two reactions that follow?

———

$$AB \longrightarrow A + B$$

$$(\Delta G' = +5{,}000 \text{ cal/mole})$$

$$B + X \longrightarrow XB$$

$$(\Delta G' = -7{,}000 \text{ cal/mole})$$

When these two reactions are coupled, the overall result is that *B* is transferred from *AB* to *XB*. Such a reaction is often written in the following notation:

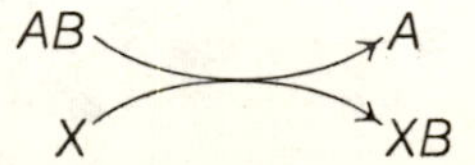

(Note that *B* is not shown as a free entity in this notation.)

➡ If you already know how to use this notation, you may jump to frame 2.27. If you cannot answer the question there correctly, return to frame 2.26.

2.26

The notation described in frame 2.25 is frequently encountered in reactions in which phosphate groups are transferred. For example, when the following reactions are coupled, a phosphate group ($H_2PO_4^-$), represented by Ⓟ, is transferred from glucose to fructose. (Glucose and fructose are represented here by highly simplified diagrams, which show only the parts of the molecule that participate in these reactions.)

2.25 *B*

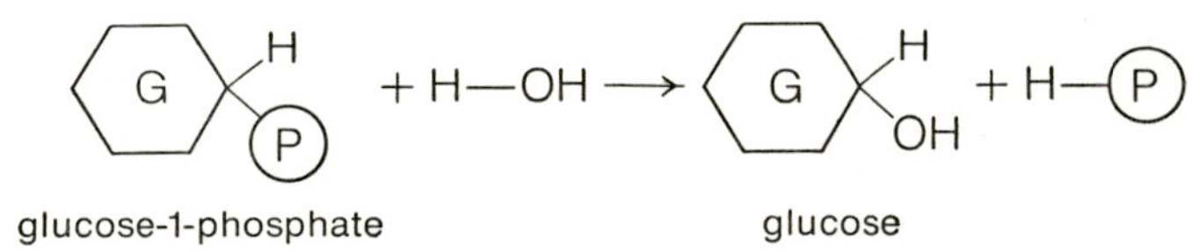

($\Delta G' = -5{,}000$ cal/mole)

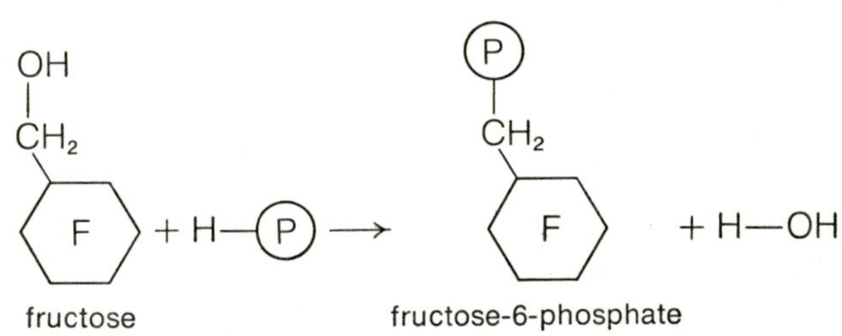

($\Delta G' = +3{,}800$ cal/mole)

In our shorthand notation these coupled reactions would be represented as either of the following:

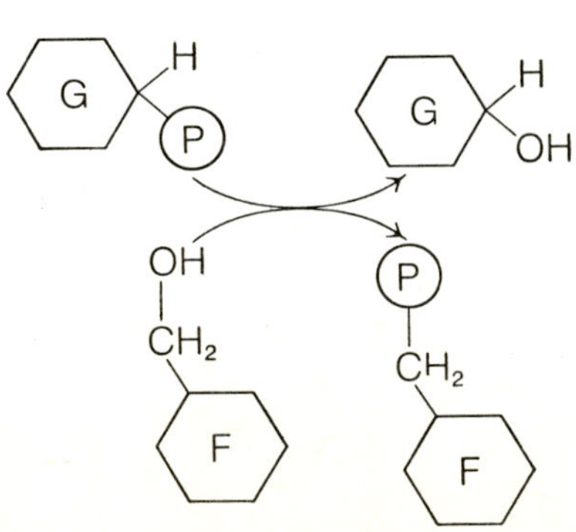

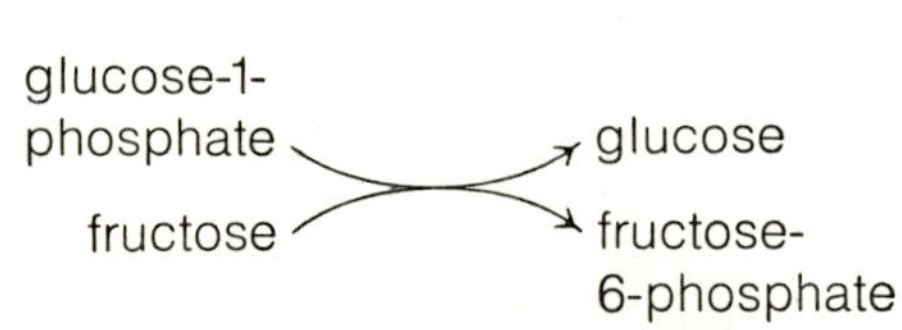

What is being transferred in the following coupled reactions? ____________

2.27 ____

2.25

Write the following pair of coupled reactions in shorthand notation:

$$\text{creatine phosphate} + H_2O \longrightarrow \text{creatine} + H_3PO_4 \quad (\Delta G' = -10{,}500 \text{ cal/mole})$$

$$\text{glycerol} + H_3PO_4 \longrightarrow \text{glycerol phosphate} + H_2O \quad (\Delta G' = +2{,}300 \text{ cal/mole})$$

2.26
glucose

2.28 ____

Notation for a series of coupled reactions There is also a shorthand notation for a long series of coupled reactions. For example, the following hypothetical series of four coupled reactions

1. $A + B \longrightarrow C + D$
2. $C + E \longrightarrow F$
3. $F \longrightarrow G + H$
4. $H \longrightarrow I$

would be written in the following way:

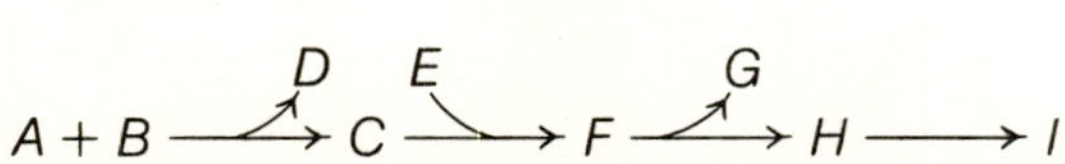

Note how D and G are shown leaving the pathway and E is shown entering the pathway.

2.27

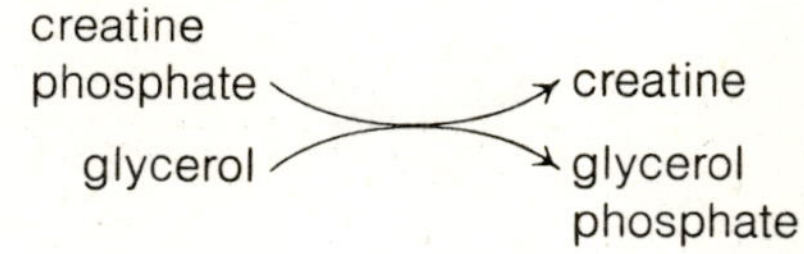

Using the shorthand notation, rewrite the following series of coupled reactions:

1. $X + Y \longrightarrow L$
2. $L + M \longrightarrow P$
3. $P \longrightarrow T + S$
4. $S \longrightarrow W + O$
5. $W + G \longrightarrow A$

2.29

Following is a series of coupled reactions that occurs in glycolysis. Write these reactions using the shorthand notation.

fructose diphosphate $\longrightarrow$ 2 glyceraldehyde phosphate

2 glyceraldehyde phosphate $+ 2\ H_3PO_4 \longrightarrow 4$ "H" + 2 diphosphoglyceric acid

2 diphosphoglyceric acid $\longrightarrow$ 2 phosphoglyceric acid $+ 2\ H_3PO_4$

2.28

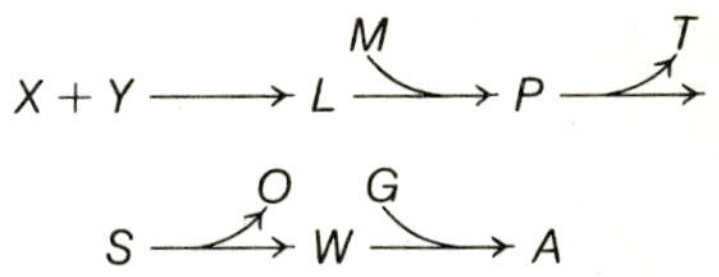

2.29

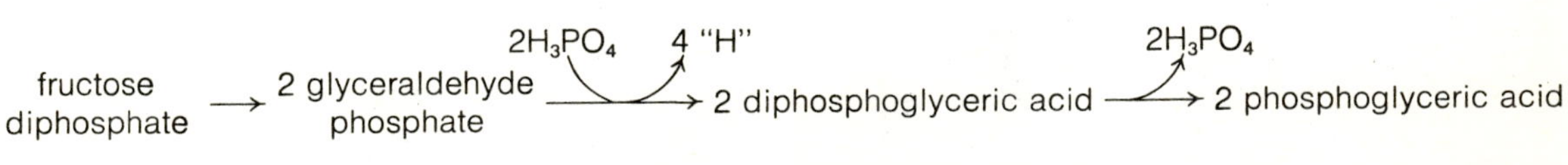

The last step could also be written as:

2 diphosphoglyceric acid $\rightarrow 2\ H_3PO_4 +$ 2 phosphoglyceric acid

four

practice problems to test your understanding of this unit on coupled reactions

2.30 ______________________________

In a pair of coupled reactions, an
____endergonic reaction
____exergonic reaction
drives an
____endergonic reaction
____exergonic reaction
in the direction of product formation.

2.31 ______________________________

What two conditions must be met for reactions to be coupled?

2.30
exergonic reaction;
endergonic reaction

2.32 ______________________________

Can the following two reactions be coupled? If not, explain why.

2.31
The reactions must share

phosphoenolpyruvic acid + H_2O ⟶ pyruvic acid + H_3PO_4

$(\Delta G' = -12{,}800 \text{ cal/mole})$

glucose + H_3PO_4 ⟶ glucose-6-phosphate + H_2O

$(\Delta G' = +3{,}300 \text{ cal/mole})$

common intermediate(s) and the sum of their ΔGs must be negative.

2.33

Use shorthand notation to write the two coupled reactions shown in the preceding frame.

2.32
yes

2.34

Explain why the endergonic reaction in frame 2.32 proceeds.

2.33

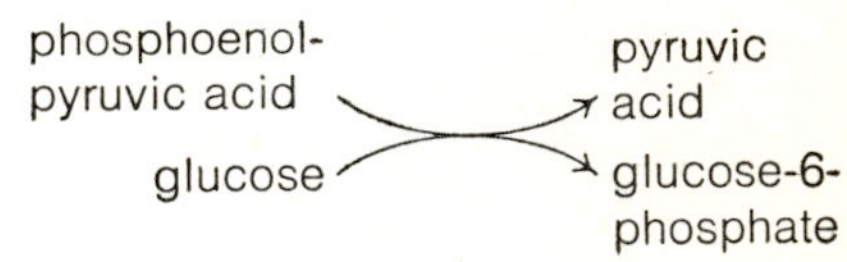

2.35

Is there a limit to the number of reactions that can be coupled? If so, what is it?

2.34
The exergonic reaction produces large amounts of H_3PO_4, which upsets the balance between reactants and products of the endergonic reaction. To compensate, the endergonic reaction proceeds.

2.36

Can the following three reactions be coupled? If not, explain why.

ethanol ⟶ 2 "H" + acetaldehyde

$(\Delta G' = +5{,}000 \text{ cal/mole})$

CO_2 + acetaldehyde ⟶ pyruvic acid

$(\Delta G' = +4{,}700 \text{ cal/mole})$

pyruvic acid + 2 "H" ⟶ lactic acid

$(\Delta G' = -6{,}000 \text{ cal/mole})$

2.35
no, there is no limit

2.37

Assume for a moment that the three reactions in frame 2.36 could be coupled and write the series in conventional shorthand notation.

2.36
no, because the sum of the ΔG's is not negative

2.38

Can the following reactions be coupled? If not, explain why.

oxalacetic acid + 2 "H" ⟶ malic acid
$(\Delta G' = -6{,}700 \text{ cal/mole})$

pyruvic acid + CO_2 + 2 "H" ⟶ malic acid
$(\Delta G' = +300 \text{ cal/mole})$

2.37

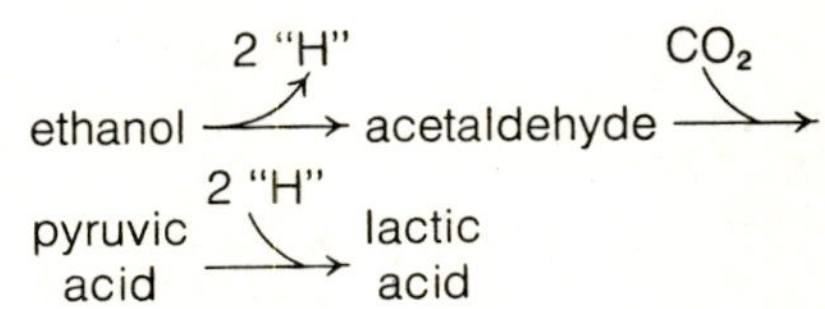

● Go to Unit Three.

help frames for unit two

2.39
2.02

The effects of adding or removing substances on a chemical system at equilibrium The ratio of products to reactants at equilibrium is a constant. When we add or remove substances from a reaction mixture that is at equilibrium, we change the ratio of products to reactants and upset the equilibrium. If we do nothing further, the reaction will return to equilibrium. Consider the following situation:

$$X \rightleftarrows Y$$

At equilibrium, the ratio of [products] to [reactants], or [Y] to [X], is 1:2. Therefore, a mixture of 3 moles of Y and 6 moles of X will be in equilibrium.

2.38
no, because the reactions do not have a common intermediate

If we add 3 more moles of X to the mixture, the ratio of products to reactants at that instant is ________________, and the system is no longer in equilibrium.

2.40 ______________________________

The system will now tend to return to equilibrium. To do so, it will have to reduce its excess of reactant. X will be converted to Y.

Adding reactant to an equilibrium mixture drives the reaction

_____forward.
_____in reverse.

2.41 ______________________________

What will be the effect of removing product from an equilibrium mixture?

2.42 ______________________________

Let us consider another reaction:

$$\text{sucrose} + H_2O \rightleftarrows \text{glucose} + \text{fructose}$$

If we add sucrose to an equilibrium mixture, ____________ ________ will be converted to ____________________ until equilibrium is reestablished.

2.43 ______________________________

If we remove sucrose from an equilibrium mixture, ________ ____________ will be converted to ____________________ until equilibrium is reestablished.

2.44 ______________________________

Frames 2.42 and 2.43 can be summarized as follows:

- In response to any change made in the concentration of reactants or products after equilibrium has been reached, the reaction proceeds in whichever direction will nullify the change and reestablish equilibrium.

In terms of the generalized reaction $A + B \rightleftarrows C + D$, this means that if we add A or B or remove C or D, the reaction is

2.39
3:9 or 1:3

2.40
forward

2.41
The reaction will be driven forward.

2.42
Sucrose and water will be converted to glucose and fructose.

2.43
Glucose and fructose will be converted to sucrose and water.

forced forward. If we add C or D or remove A or B, the reaction is forced in reverse.

Return to frame 2.02.

2.44
no answer

unit three

rate of reaction

one
introduction

3.01

As you have learned thus far, many reactions essential to life do not proceed spontaneously (that is, they have a positive $\Delta G'$). In the living cell this obstacle is overcome by coupling the nonspontaneous reactions to exergonic reactions with which they share common intermediates. Under such conditions the nonspontaneous reactions do proceed. At this point it is appropriate to ask whether they proceed fast enough to meet the needs of living things.

3.02

Coupled reactions have a net negative ΔG.

Does a net negative ΔG imply that the reaction proceeds at a relatively rapid rate?

____yes

____no

3.01
no answer

3.03

A reaction with a large negative ΔG does not necessarily occur at a rapid or even an observable rate, as this familiar example demonstrates:

$$\text{sucrose (solid)} + 12\ O_2 \longrightarrow 12\ CO_2 + 11\ H_2O \quad (\Delta G^\circ = -1{,}363{,}000\ \text{cal/mole})$$

Sucrose is ordinary table sugar. The large negative ΔG° tells us that sucrose spontaneously combines with oxygen at one atmosphere pressure and room temperature. Yet sugar can remain in an open dish for months without undergoing any noticeable changes because the rate of reaction is exceedingly slow.

- The ΔG of a reaction indicates the direction of spontaneous change, that is, whether the concentration of products or the concentration of reactants will increase spontaneously. However, ΔG does not indicate how fast the spontaneous change will take place.

3.02
no

➡ If you answered correctly, go to frame 3.04. If you did not, go to frame 3.03.

3.04
3.02

Most reactions that normally occur in cells proceed only very slowly outside the cell.

Example

The contraction of muscle depends on the following exergonic reaction:

$$ATP + H_2O \longrightarrow ADP + H_3PO_4$$

When this reaction occurs in a test tube at 98.6° F and pH = 7, it is completed in approximately four weeks.

There must be some means by which the cell can accelerate reactions, otherwise life would not be possible. If the preceding reaction were not accelerated, the organism could not move fast enough to escape unfavorable changes in its environment. Our deduction that reactions are somehow accelerated in the cell is supported by observations that inside the cell this reaction is completed in two seconds.

The means by which slow reactions are accelerated in the cell will be the major topic of this unit. First we will investigate what happens as molecules of reactants interact to form molecules of product.

3.03
no answer

3.04
no answer

two
a chemical reaction seen at the molecular level

3.05

This frame summarizes the events that must occur during a reaction.

First, molecules of reactants must come into contact by colliding with each other. Often the molecules must encounter each other in a certain spatial orientation. For example, in the following reaction

$$Cl^- + CH_3OH \longrightarrow CH_3Cl + OH^-$$

the Cl^- has to attack opposite to the $-OH$ group; otherwise it will be repelled by the electron shells of the $-OH$ group.

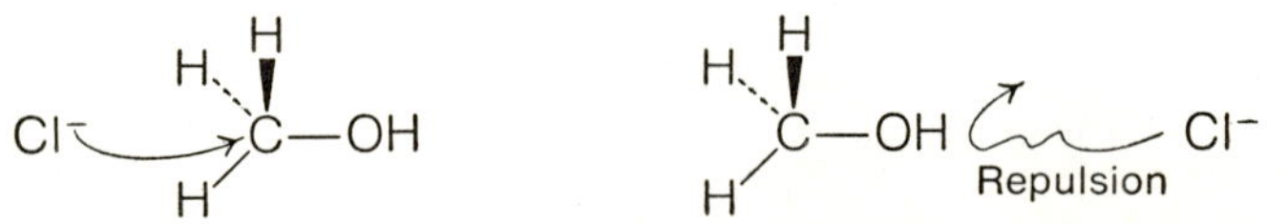

Successful attack   Unsuccessful attack

Second, existing bonds in the reactants must be broken or disrupted. When bonds in a molecule are disrupted, atoms are separated. Such a displacement of atoms increases the potential energy of the molecule, as shown in the following diagram:

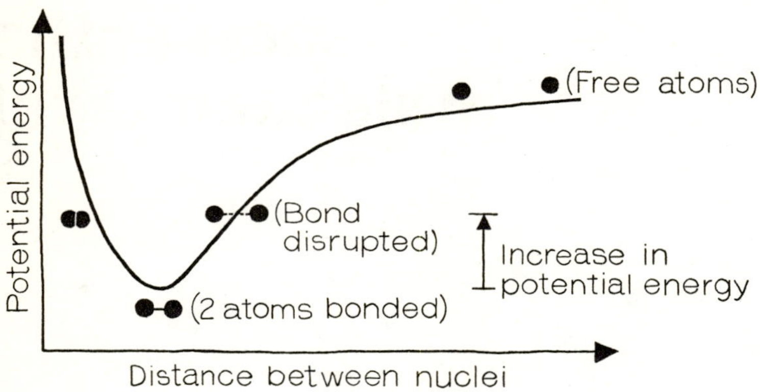

In addition, new bonds must be formed. Formation of new bonds can be accompanied by an increase in potential energy (for example, when like charges must be brought together).

- The configuration of atoms that corresponds to the highest potential energy during the course of a reaction is called the transition state.

➡ If the material in this frame has been a review of what you already knew, skip to frame 3.07.

3.06

Here are two reactions and their transition states:*

3.05
no answer

a.

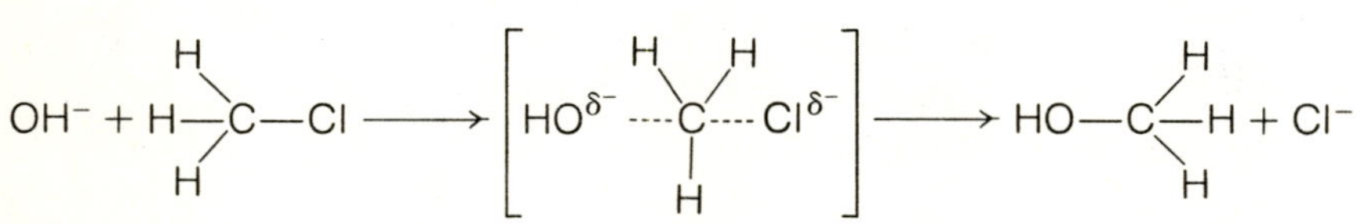

b.

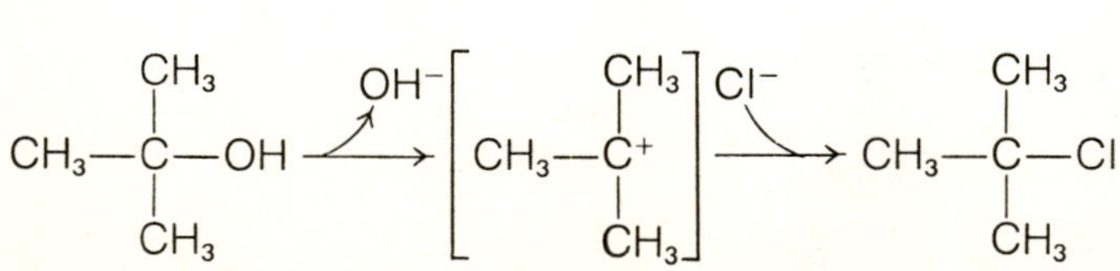

*---- = partial bond
δ^- = partial negative charge

In what order do the essential events described in frame 3.05 occur in each of the reactions?

a. ____

____

b. ____

____

3.07 ____

3.05

Let us consider how the events that occur during a reaction would affect the disorder of a system containing reactants at a low concentration. The following system provides a fair model of cytoplasm:

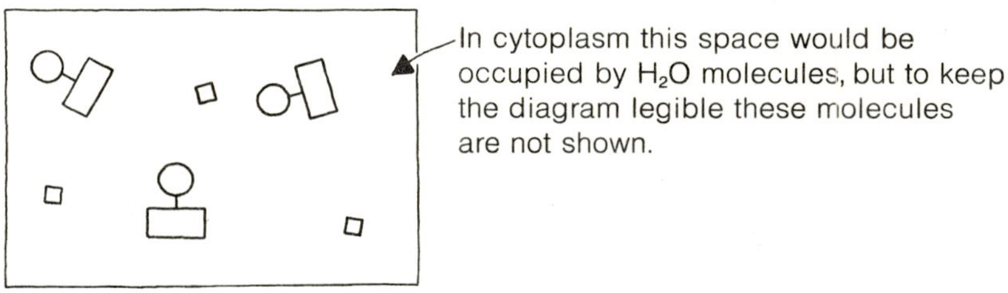

Model reaction to be considered: 

The events that must occur during a reaction were described in frame 3.05. These requirements may be expressed in terms of our model system as follows:

Requirement 1. Molecules of reactants come into contact.

3.06

a. oriented collision, simultaneous disruption of bond and formation of new bond

b. breakage of bond, oriented collision, and formation of new bond (The first step of this reaction was simplified for pedagogical reasons. Actually, the transition state does not form unless a proton is present, and H_2O rather than OH^- leaves the reactant molecule.)

a. Disorder of the system has
____increased.
____decreased.

Requirement 2. Molecules of reactants collide with proper orientation.

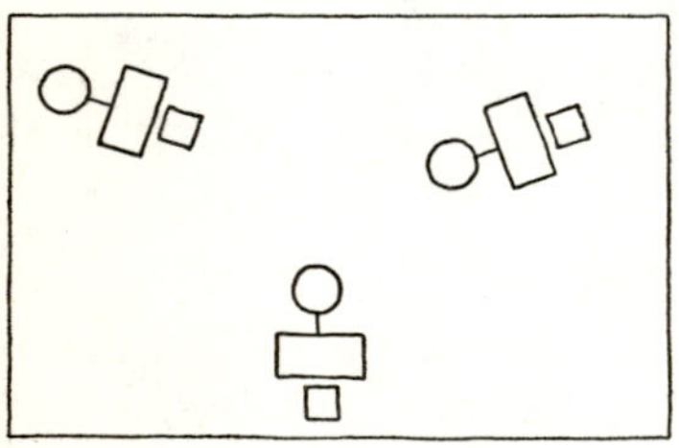

b. Disorder of the system has
____increased.
____decreased.

Requirement 3. Existing bonds in the reactants are broken or disrupted and new bonds are formed.

c. This process leads to the formation of the
________________________.

As the transition state is formed, part of the translational, vibrational, and rotational kinetic energies of reactant molecules must be converted to potential energy of the transition state. At the temperatures characteristic of living cells, the kinetic energies of most molecules are relatively low, and the amount of kinetic energy that must be converted to potential energy is a significant part of the total kinetic energy:

$$\left.\begin{array}{l} K.E._{\text{translational}} \\ K.E._{\text{vibrational}} \\ K.E._{\text{rotational}} \end{array}\right\} \rightarrow \text{Potential energy of transition state}$$

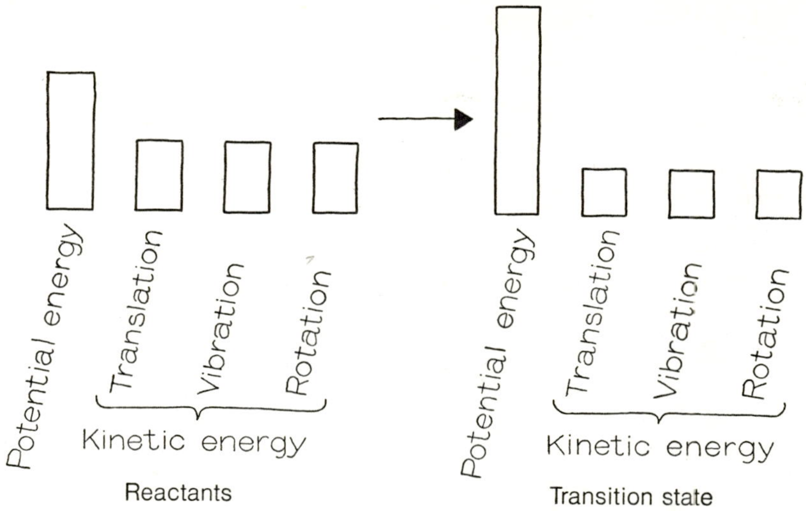

d. When the transition state is formed, the distribution of energy becomes
____more even.
____less even.

e. Therefore, the change in energy distribution required for disruption of old bonds and formation of new bonds will
____increase the disorder of the system.
____decrease the disorder of the system.

3.08

To summarize, our model shows that each step in the formation of the transition state from average reactant molecules decreases the disorder of the system.

Therefore, the probability that each step will occur is
____high.
____average.
____low.

3.07
a. decreased
b. decreased
c. transition state
d. less even
e. decrease the disorder of the system

3.09

What will be the rate of reaction in the model system? We know from the way energy is distributed in any collection of molecules that the kinetic energies of a few molecules in the system will be much higher than average.

3.08
low

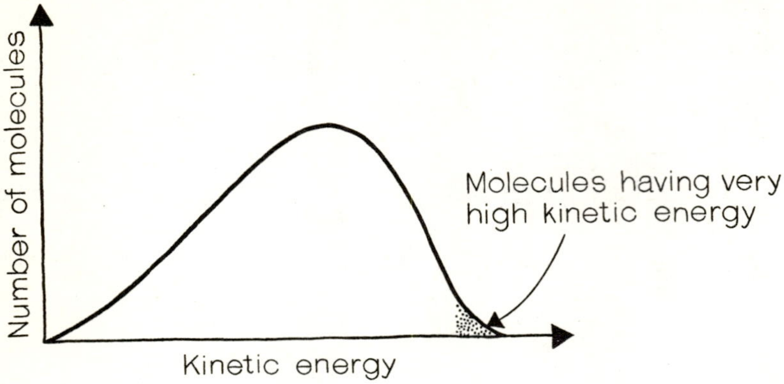

For these molecules, conversion of some kinetic energy into potential energy of the transition state does not result in decreased disorder. Therefore, when these high-energy molecules collide in the proper orientation they will probably form the transition state and then convert to products. However, this will be an extremely rare event because properly oriented collisions represent only a minute fraction of the total collisions taking place, and high-energy molecules are only a tiny proportion of the total molecules in the system.

Thus, the rate of reaction in the model system will be

____fast.
____average.
____slow.

3.10

Free energy terminology As you saw in frame 3.07, each step in the formation of the transition state from average reactants causes a decrease in disorder of the model system. We can express this decrease in disorder as a change in free energy (ΔG).

3.09
slow

- The ΔG for the conversion of reactants into the transition state is called the free energy of activation.

For our model system, the free energy of activation will be

____positive.
____negative.

3.11 ____

The following diagram shows the changes in free energy for a typical exergonic reaction in the cell.

3.10
positive (because the disorder of the system decreases as the transition state is formed)

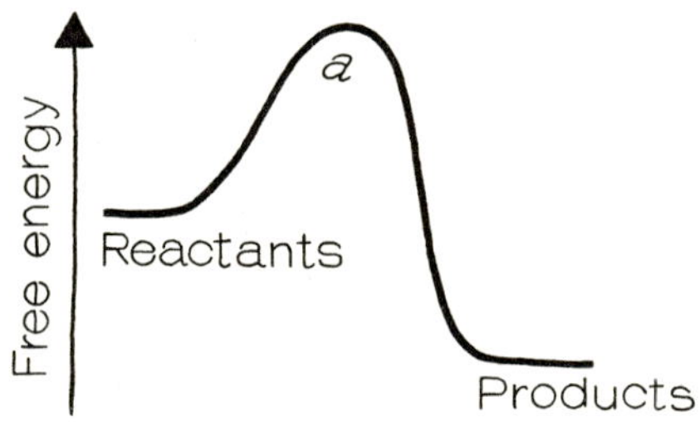

The configuration of atoms that corresponds to the point on the reaction pathway labeled *a* is called the

______________________.

● For help, see frame 3.44.

3.12 ____

On the preceding diagram, draw the line that represents the free energy of activation.

3.11
transition state

● Go to frame 3.44 if you missed this question.

summary

3.13 ____

The steps necessary for formation of product(s) from reactants are:

1. Collision between reactant molecules. This step decreases the disorder of the system.
2. Disruption or breakage of existing bonds and formation of new bonds. In this step, kinetic energy is converted to potential energy. At temperatures characteristic of the living cell, such energy conversion decreases the disorder of the system.

3.12

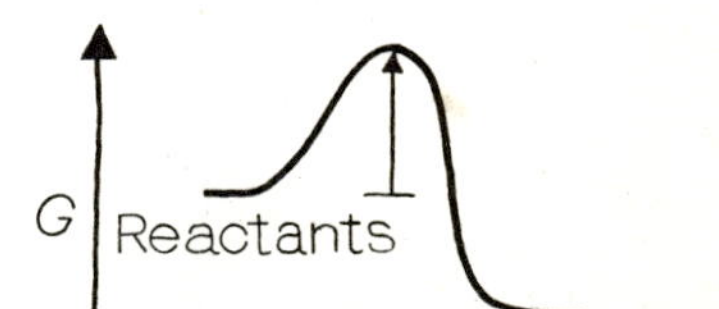

Free energy of activation = distance between reactants and top of curve

The configuration of atoms that corresponds to the maximum potential energy during the course of a reaction is called the transition state. The ΔG for formation of the transition state from reactants is called the free energy of activation. Because both steps in the formation of the transition state decrease the disorder of the system, the free energy of activation is positive.

3.13
no answer

three

accelerating reactions: nonbiological methods

3.14

Increase in temperature A common method of hastening slow reactions in the laboratory is to heat the reaction mixture. Of course, heating is not a practical way to increase the rates of reactions in living organisms since excessive heat damages tissue. However, it is instructive to consider how a rise in temperature affects each of the necessary steps in a reaction.

Can you recall what these necessary steps are?

3.15

When a collection of molecules is heated, the kinetic energies of the molecules are affected. As the following diagram shows, when the temperature is raised the most probable kinetic energy is higher. In addition, the proportion of molecules

3.14
Reactant molecules must come into contact (collide). Often the collision must

with above-average kinetic energies increases. The greater its kinetic energy, the faster a molecule moves about.

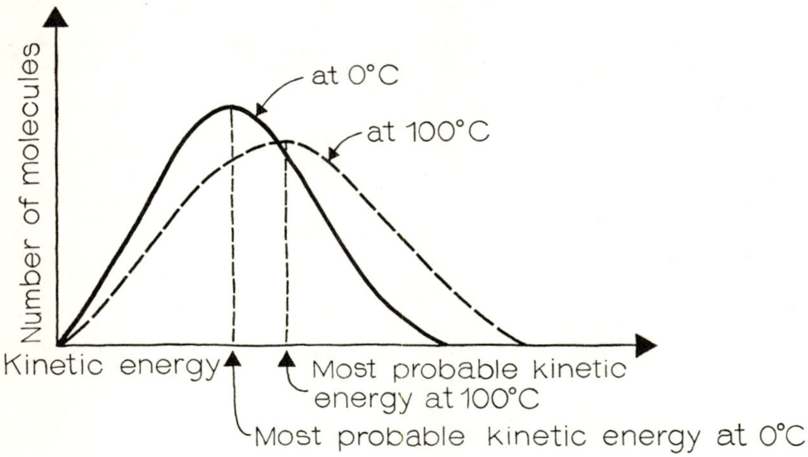

occur with a certain mutual orientation to be effective. Existing bonds must be broken or disrupted and new bonds must be formed. This requires the conversion of kinetic energy (translational, vibrational, and rotational) to potential energy.

3.16

What effect will a temperature increase have on the probability of collisions?

3.15
no answer

● If you need help, go to frame 3.45.

3.17

Would you expect an increase in temperature to enhance the probability that collisions between molecules will occur with the proper mutual orientation? Why or why not?

3.16
More collisions should occur per unit of time, collisions become more probable, and so on.

3.18

At temperatures characteristic of living cells, conversion of some kinetic energy of reactants to potential energy of the transition state makes the distribution of energy less even and decreases the disorder of the system. At higher temperatures, molecules have more kinetic energy.

3.17
Temperature increase should not exert any effect other than simply increasing the rate of collisions.

Conversion of some of this kinetic energy to potential energy will cause a decrease in disorder that is
____larger
____smaller
than the decrease in disorder that occurs at the lower temperature.

3.19 ____

If two reactant molecules have collided in the proper orientation at 40° C and another two molecules have collided in the same orientation at 300° C, which pair of molecules is more likely to undergo the redistribution of energy necessary to disrupt bonds in the reactants and form the transition state?
____molecules at 40° C
____molecules at 300° C

3.18
smaller

3.20 ____

To summarize, increasing the temperature of a reaction mixture accelerates the reaction for the following reasons:

1. The number of collisions per unit of time is increased because the molecules are moving faster at the higher temperature.

2. The molecules have higher kinetic energy, so conversion of some of the kinetic energy to potential energy of the transition state causes a smaller decrease in disorder of energy distribution. Thus, two collided reactants are more likely to form the transition state at a higher temperature.

3.19
molecules at 300° C

3.21 ____

Catalysis There are many reactions that can be accelerated by adding a particular substance, such as a metal ion, an acid, or a base. For example, the reaction

$$\text{ATP} + H_2O \longrightarrow \text{ADP} + H_3PO_4$$

proceeds approximately ten times faster when a small amount of acid is added to the reaction mixture.

The substance that enhances the rate is always present at the end of the reaction in the same concentration as it was at the start of the reaction, so it is not consumed or permanently changed by the reaction. Such a substance is called a catalyst.

3.20
no answer

3.22

How catalysts work Catalysts enhance reaction rates by lowering the difference in free energy between the reactants and the transition state. They can accomplish this in two ways:

3.21
no answer

1. By increasing the stability of the transition state.
2. By decreasing the stability of the reactants. This can be done by weakening the bond(s) that must be broken or disrupted during the formation of the transition state, by increasing the reactivity of an attacking molecule, and so on.

Both of these strategies reduce the difference in potential energy between reactants and transition state. Therefore, less kinetic energy has to be converted to potential energy during the formation of the transition state, and disorder of energy distribution decreases less. This effect lowers the free energy of activation.

However, we must keep in mind that collisions between three particles (two reactants and one catalyst) will cost more in terms of decreased positional disorder than collisions between two particles (reactants only). This decrease in positional disorder is a side effect that somewhat counteracts the energy distribution advantage of a catalyst, but the net effect of a catalyst is to lower the free energy of activation. In other words, although encounters between catalyst and reactants are rarer than encounters between reactants, the former encounters are much more likely to lead to formation of the transition state when they do occur.

The following diagram represents free energy changes during an uncatalyzed reaction. Draw a similar diagram to

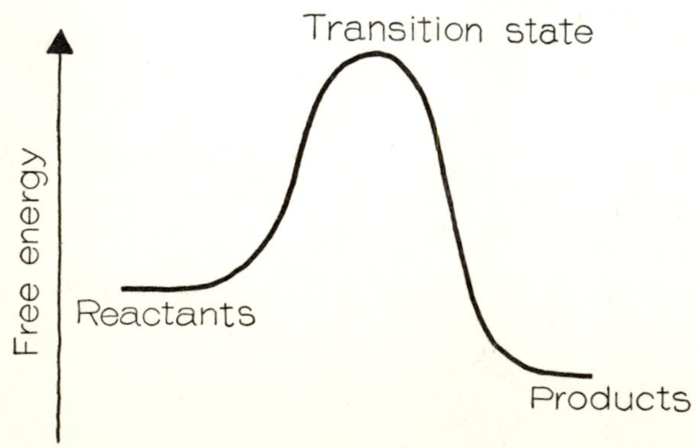

show the free energy changes that would result if the same reaction took place in the presence of a catalyst.

3.23

Examples of catalytic mechanisms

1. Catalysis achieved by increasing the stability of the transition state.

The following transition state†

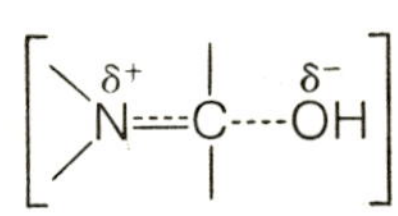

is stabilized by an acid, represented by A-H:

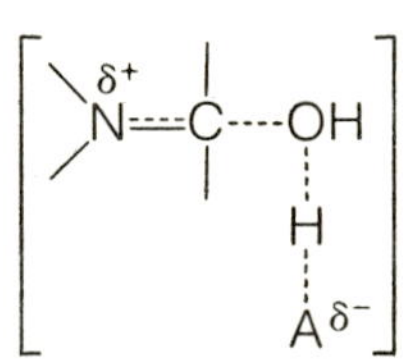

An acid (A-H) is a substance that gives up a proton more readily than water does because A^- is more stable than OH^-. The greater stability of A^- implies that the transition state having a partial negative charge on A will be more stable than the transition state having a partial negative charge on OH.

2. Catalysis achieved by decreasing the stability of the reactants.

† ---- = partial bond
δ^- = partial negative charge

3.22

In your diagram, reactants and products should occupy the same positions they occupied in the diagram of the uncatalyzed reaction but the transition state should be lower.

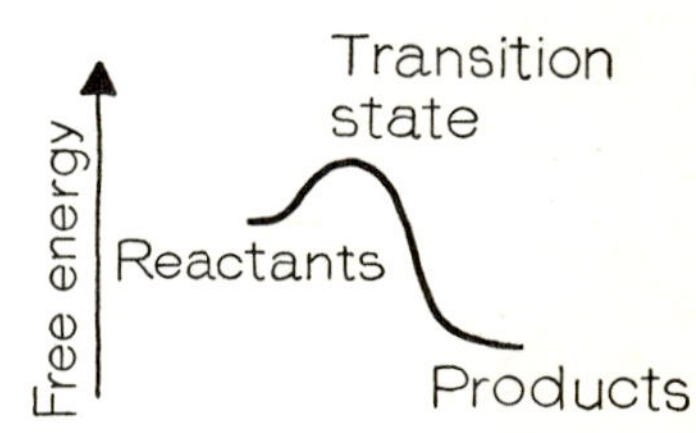

The following reaction

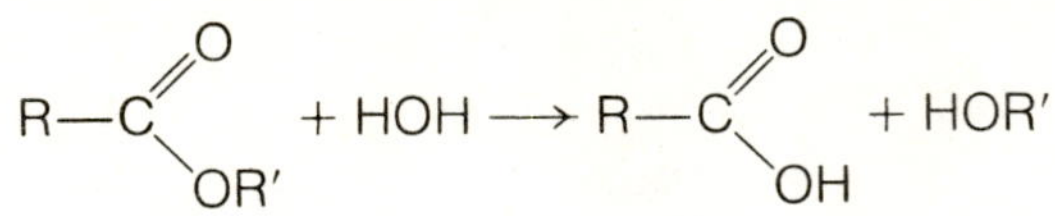

(ester)

is catalyzed by acids. The acid donates a proton to the ester:

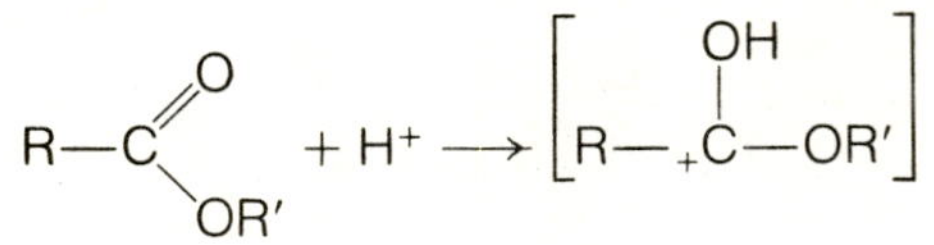

This results in a rearrangement of electrons that leaves the carbon atom deficient in electrons and therefore positively charged. The carbon atom is now much more susceptible to attack by the oxygen atom of a water molecule, because the electron-rich oxygen atom can approach more closely without being repelled by electrons of the carbon atom.

3.24 ____

In the preceding frame you saw two examples in which addition of an acid can accelerate a reaction. Could the reaction rates be enhanced significantly by a small amount of acid (say, one molecule of acid per 1,000 molecules of reactants), or would nearly equal concentrations of acid and reactants be required to produce a significant increase in the reaction rates? Briefly explain your reasoning.

3.23
no answer

● For help, see frame 3.46.

3.25 ____

What effect, if any, does a catalyst have on the free energy of activation?

3.24
A very low concentration of catalyst can speed up a reaction. Reason: One molecule of catalyst can be used over and over, since it undergoes no permanent change and is always regenerated by the reaction it catalyzes.

3.26

What effect, if any, does a catalyst have on the ΔG of a reaction?

3.25
A catalyst lowers or reduces the free energy of activation.

3.27

Decide whether each of the interactions described in this frame is an example of catalysis. (Note: These examples and chemical structures are provided for students with a strong background and interest in chemistry. If they are confusing to you, ignore them and concentrate on the written description of each interaction.)

1. Increasing the reactivity of an attacking group:

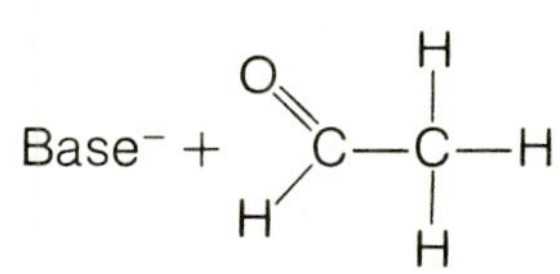

If a base removes a proton from the molecule on the right, the resulting negative ion is more likely to react with carbon atoms having a relative deficiency of electrons:

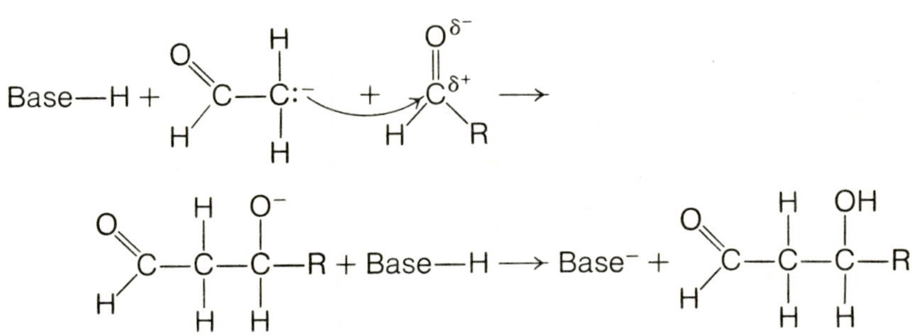

a. Example of catalysis?
_____yes
_____no

2. Reducing the number of like charges that must be brought together during a reaction:

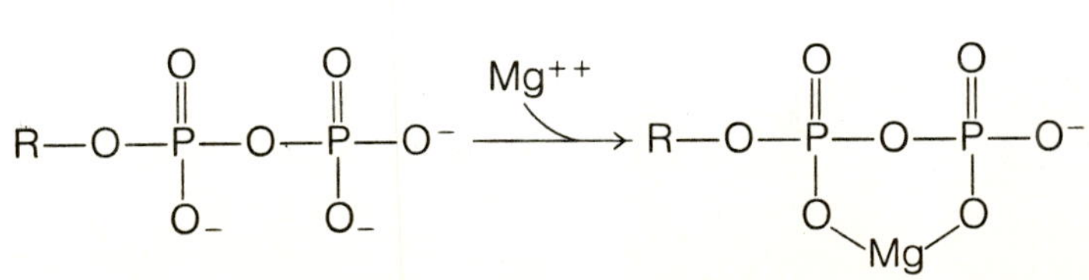

3.26
none

● If you missed this question, see frame 3.47.

If a metal ion such as Mg^{++} neutralizes two of the negative charges on a molecule containing a diphosphate group, the electrons of an attacking water molecule can approach more easily:

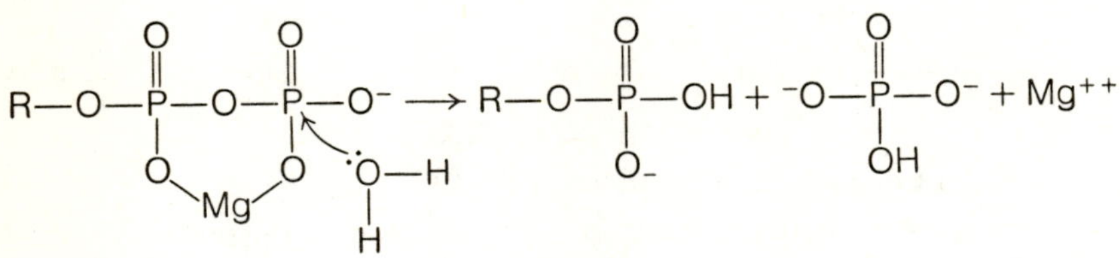

b. Example of catalysis?
____yes
____no

3. Increasing the stability of an attacking group.

Original reaction:

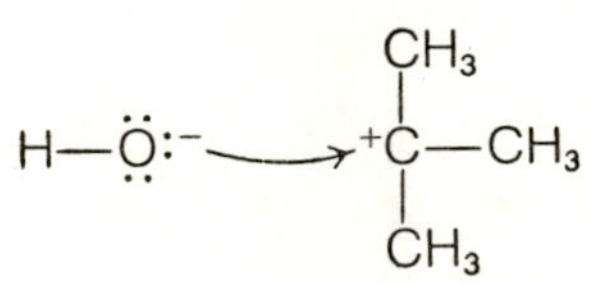

Reaction after the stability of OH^- has been increased by adding a proton:

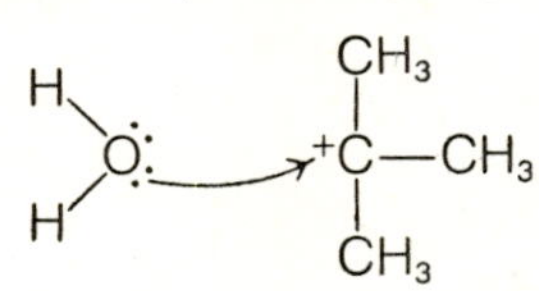

c. Example of catalysis?
____yes
____no

4. Weakening the bond between the main part of the molecule and a group that leaves during the reaction.

Original reaction:

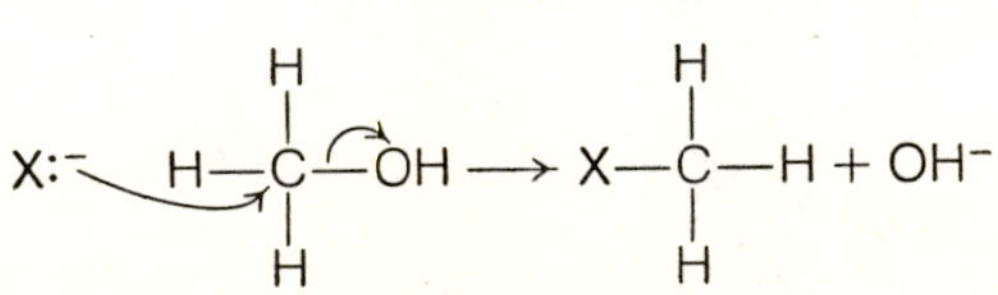

R o in presence of acid:

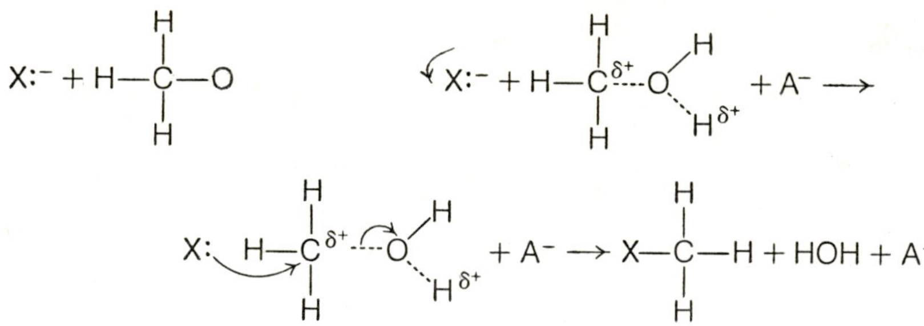

d. Example of catalysis?

_____yes

_____no

3.27

a. yes
b. yes
c. no
d. yes

summary

3.28

A reaction can be accelerated by raising the temperature of the reaction mixture or by adding a catalyst. Raising the temperature increases the kinetic energies of the molecules. Thus more collisions occur per unit of time because the molecules are moving faster, and in many of the collisions the conversion of kinetic energy to potential energy causes a smaller decrease in disorder. A catalyst is a substance that accelerates a reaction without being consumed or permanently changed by the reaction. Catalysts lower the free energy of activation either by increasing the stability of the transition state or decreasing the stability of the reactants.

3.28

no answer

four
how the cell accelerates reactions

3.29

Cells manufacture a type of catalyst that is even more effective in increasing reaction rates than are the catalysts that were discussed in the preceding section. These cellular catalysts are called enzymes.

Enzyme structure

All the enzymes that have been discovered are proteins, which are composed of one or more polypeptides (long chains of amino acids joined together). The polypeptide chains become folded in a complex manner as a result of interactions among the individual amino acids. This folding results in a globular molecule with a cleft or hole on its surface:

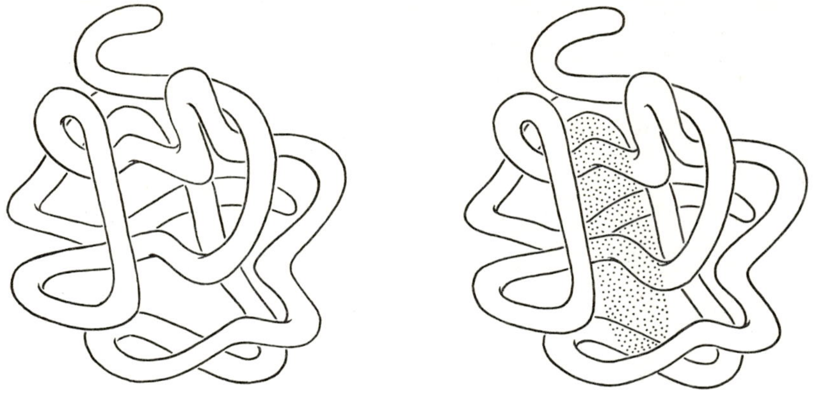

Folding of a polypeptide chain. Dotted area indicates location of cleft.

Bound to the protein of many enzymes are metal ions or small organic molecules. These nonprotein portions of the enzyme are not located at random in the enzyme but rather occupy a definite position, which varies with the type of enzyme and type of ion or small molecule.

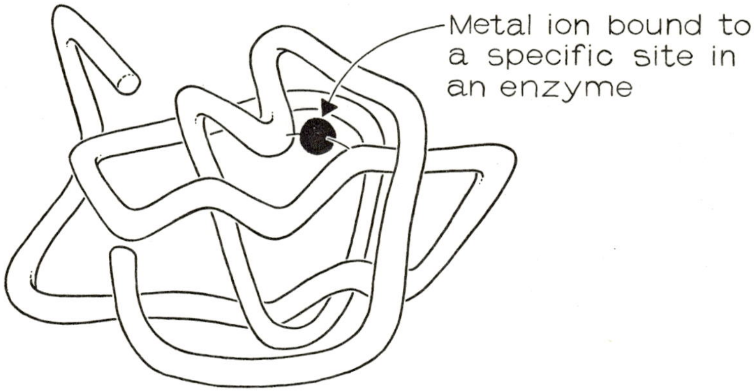

Specificity of enzymes

A notable feature of enzymes is their high degree of specificity. Whereas an ordinary catalyst such as H^+ can accelerate a wide variety of reactions, a particular enzyme usually catalyzes only one reaction or a group of very similar reactions. The reason that enzymes are so specific will become clear when we consider the mechanism of enzyme action.

review question

3.30 ___

Enzymes are catalysts. Therefore, which of the following statements should be true of enzymes?

An enzyme

____works by increasing the kinetic energies of the reactant molecules.

____is not used up or permanently altered by the reaction it catalyzes.

____lowers the ΔG of the reaction.

3.29
no answer

3.31 ___

Mechanism of enzyme action Following are two specialized terms that will be used in this discussion.

- The reactants of an enzyme-catalyzed reaction are also called the substrates of the enzyme.

  The cleft on the surface of the enzyme is called the active site.

3.30
is not used up or permanently altered by the reaction it catalyzes

The first step in an enzyme-catalyzed reaction is encounter between enzyme and reactant (substrate) molecules. The cleft in an enzyme is just the right size and shape to fit molecules of its particular substrate and has the right charged or noncharged groups to complement the substrate. Note that the enzyme in the following diagram is relatively much larger than the substrate.

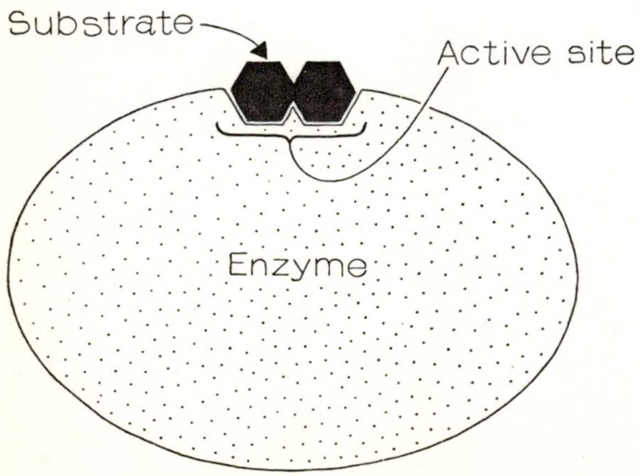

Nonsubstrate molecules do not fit well into the active site of the enzyme because they have the wrong shape or charges. The requirement of complementarity in shape between the active site and the substrate is responsible for the observed specificity of enzymes.

The active site not only conforms to the shape of the substrate but also contains functional groups of just the right sort, and in just the right positions, to form bonds of various types with the substrate. The resulting product is called the enzyme-substrate complex. Binding makes the formation of the enzyme-substrate complex a spontaneous process.

● If you are interested in the reason for this, see frame 3.48.

As the substrate fits into and binds to the active site, it comes into alignment with one or more catalytic groups that are also located in the active site. Catalytic groups in enzymes include

1. The side chains of certain amino acids that make up the protein component of the enzyme.
2. Metal ions bound to the enzyme.
3. Small organic molecules bound to the enzyme.

These groups act in the same way that nonenzyme catalysts do; that is, they
____increase the stability of the reactants.
____decrease the stability of the reactants.
____increase the stability of the transition state.
____decrease the stability of the transition state.
____increase the stability of the products.
____decrease the stability of the products.

3.32

Hypothetical example of enzyme action The following reaction

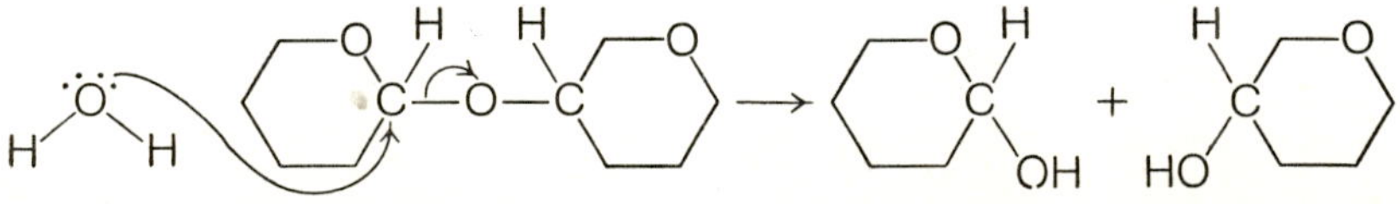

might be accelerated by an enzyme containing two catalytic groups, an acid (A^+) and a base (B^-).

3.31
decrease the stability of the reactants and increase the stability of the transition state

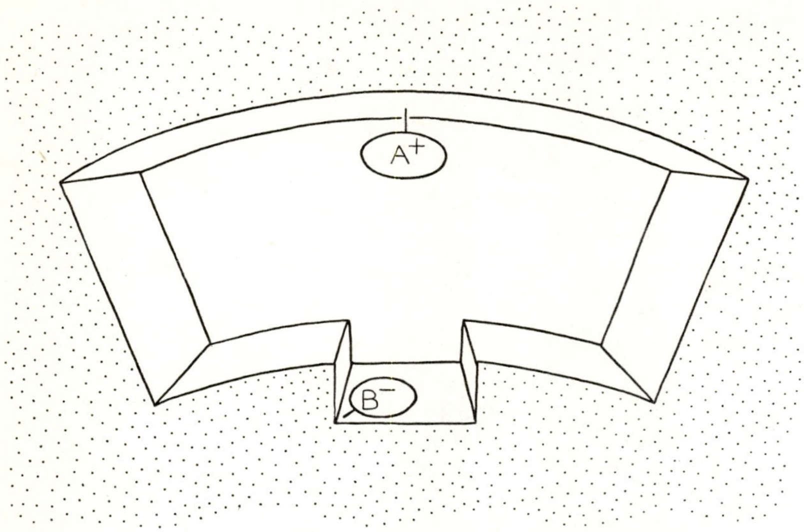

Top view, looking down on active site

The acid weakens a C—O bond by attracting some of the oxygen's bonding electrons:

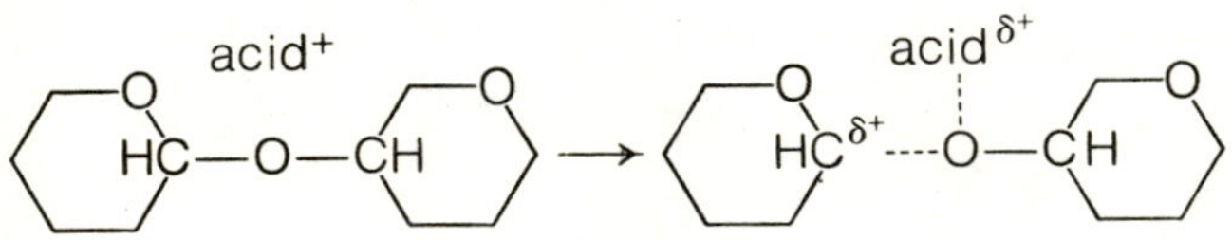

The base enhances the reactivity of water toward positively charged groups by attracting one of the water molecule's protons:

$$\text{Base:}^- + H_2O \longrightarrow \text{Base—H} + \text{:OH}^-$$

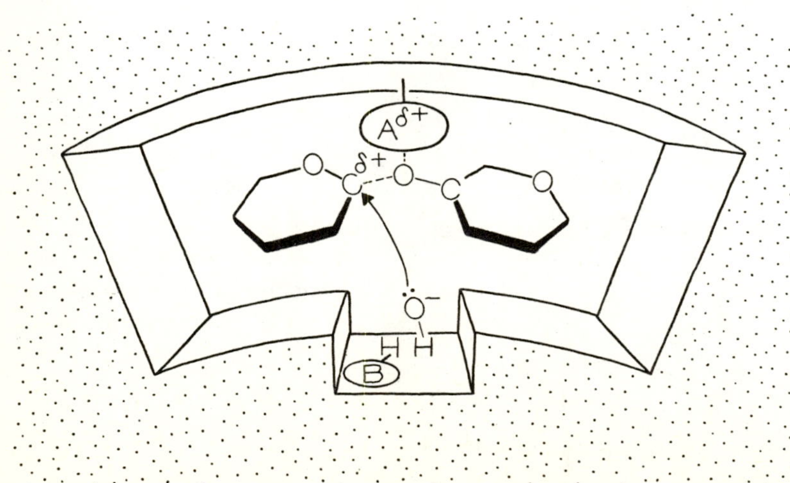

Postulated catalytic mechanism

In this hypothetical example, the model of the active site shows two essential features of a real active site, namely,

1. shape that is complementary to the substrate.
2. oriented catalytic groups.

However, the model is lacking a third essential feature of an active site. What is this feature?

3.32
groups that form bonds with the substrate, leading to formation of the enzyme-substrate complex

3.33

The great advantage of an enzyme over a nonenzyme catalyst is that the enzyme's catalytic groups are located at precisely the right places in the active site. Thus the spontaneous formation of the enzyme-substrate complex brings the catalytic groups together with the substrate in optimal mutual orientation. Catalysis by enzymes does not depend on improbable mutual collisions between reactants and one or more catalysts.

Therefore, free energies of activation should be
____higher
____lower
for enzyme-catalyzed reactions than they are for reactions that are accelerated by ordinary catalysts.

3.33
lower

3.34

Providing an optimally oriented array of catalytic groups is thought to be only one of the ways by which enzymes can accelerate reactions. Two alternative mechanisms by which enzymes could accelerate reactions are described in the following paragraphs. It is thought that some enzymes employ more than one mechanism.

1. *Alternative pathway.* The enzyme can enable a reaction to occur via a different, faster pathway than is followed during nonenzymatic reaction.

For example, consider the following reaction, in which water attacks an ester and expels phosphate ion:

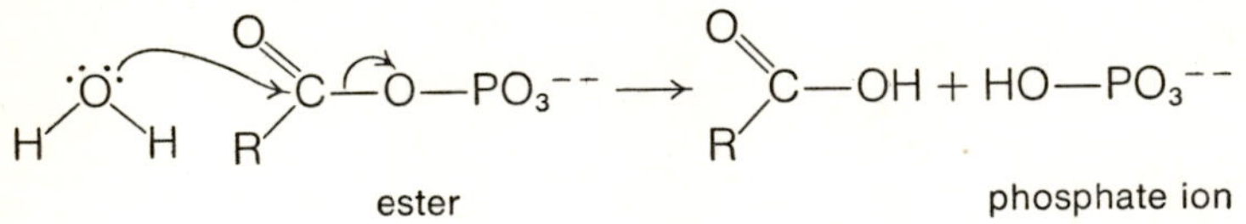

When the reaction is catalyzed by an enzyme, it proceeds as follows. After formation of the enzyme-substrate complex, the enzyme itself (by means of a sulfur atom) attacks the ester and expels phosphate ion:

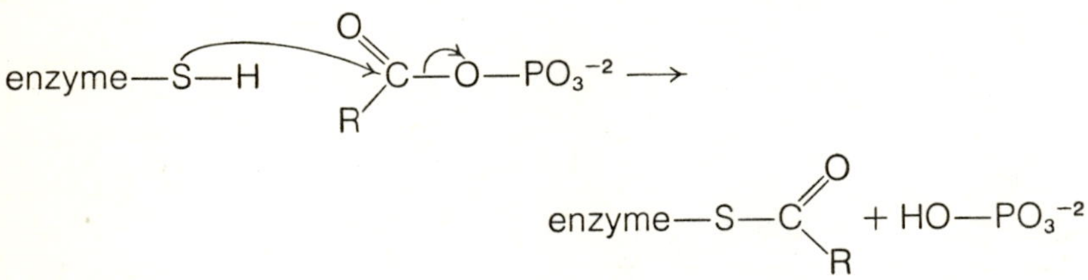

The product of this step is an enzyme-containing intermediate,

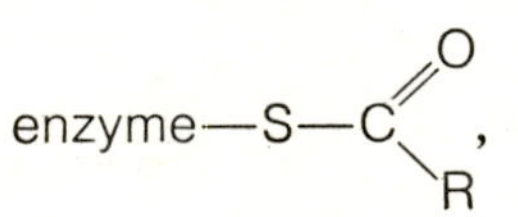

which is subsequently attacked by water:

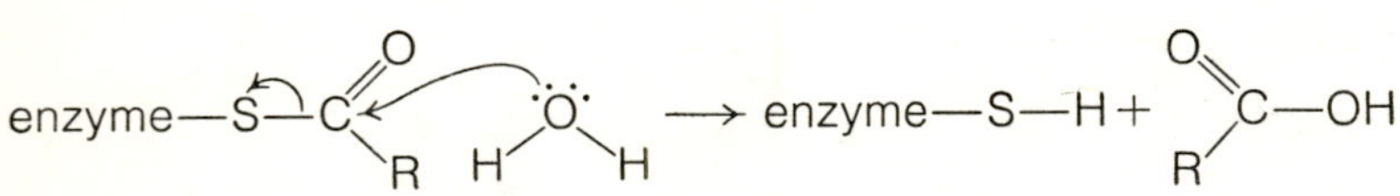

This reaction breaks up the enzyme-containing intermediate and regenerates the enzyme.

The free energy changes in the uncatalyzed and enzyme-catalyzed reactions are shown at the top of the next page. Note that the enzyme-catalyzed reaction has more steps, each of which has a lower free energy of activation than the single step in the uncatalyzed reaction. The enzyme-catalyzed reaction proceeds faster because its free energies of activation are lower than the free energy of activation of the uncatalyzed reaction.

2. *Distortion of substrate.* As the substrate and enzyme meet and bonds begin to form between them, the shape of the substrate may change as a result of binding (see illustration at

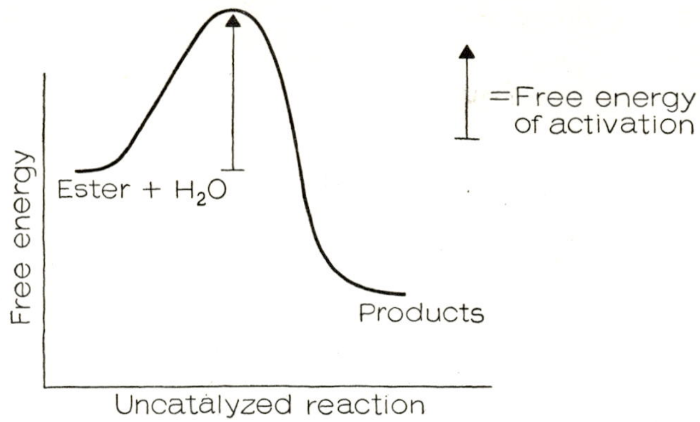

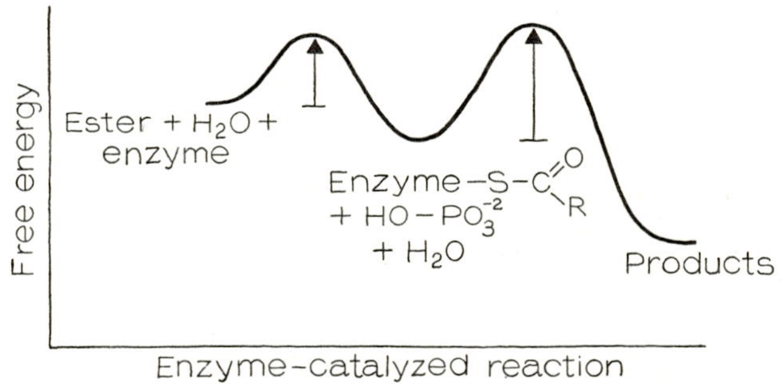

the bottom of this page). Such changes in shape of the substrate may involve stretching and weakening of bonds that are going to be broken during the course of the reaction.

In stage 2 of this hypothetical example, a bond has been stretched and thus weakened:

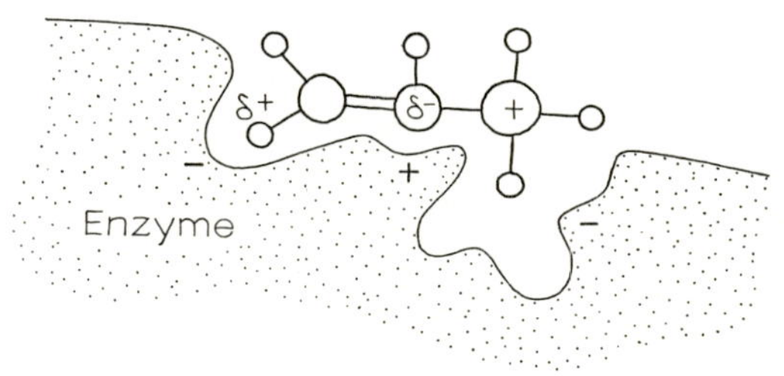

Stage 1

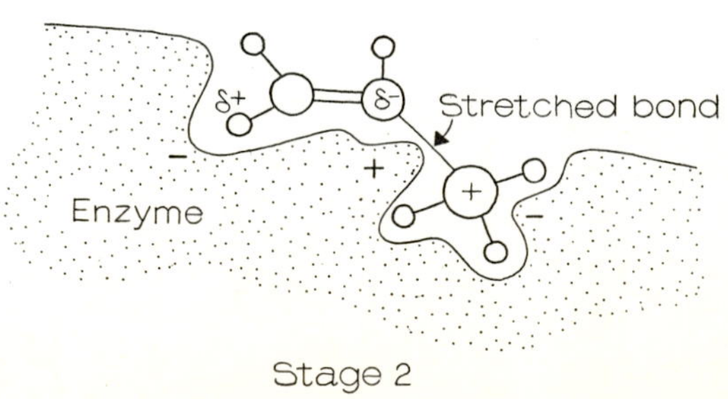

Stage 2

Note: When catalysis occurs by this mechanism, the shape of the active site will not be entirely complementary to the substrate. Instead, it will bear some resemblance to the transition state.

This method of catalysis could be classed as
____increasing the stability of the transition state.
____decreasing the stability of the reactants.

3.35

Of the catalysts that have been discussed thus far, enzymes are probably the only ones capable of catalysis by distortion. Briefly explain why.

3.34
decreasing the stability of the reactants (because a bond in the substrate is stretched and weakened)

3.36

A biologist observes a certain chemical reaction occurring in organisms but observes no such reaction when he tries to study it in a test tube. Which, if any, of the following could be the reason?

a. ____The reaction is spontaneous but proceeds so slowly without the appropriate enzyme that it does not appear to be taking place.

b. ____The reaction is endergonic. In an organism, it is coupled to an exergonic reaction that is missing in the test tube.

c. ____The ΔG is negative in an organism, but the concentrations of reactants in the test tube differ from those found in the organism, making the ΔG of the reaction positive in the test tube.

d. ____The reaction is exergonic. In an organism, it is coupled to an endergonic reaction that is not present in the test tube system.

3.35
Catalysis by distortion cannot occur with simple catalysts because it requires a site with a certain shape and with appropriate binding groups in the right places. Enzymes are the only catalysts discussed here that have a structure complex enough to fulfill these requirements.

3.37 ____

Can an enzyme make a reaction with a positive ΔG proceed?
____yes
____no

3.38 ____

Which of the following diagrams expresses the true relationship between enzymes and catalysts?

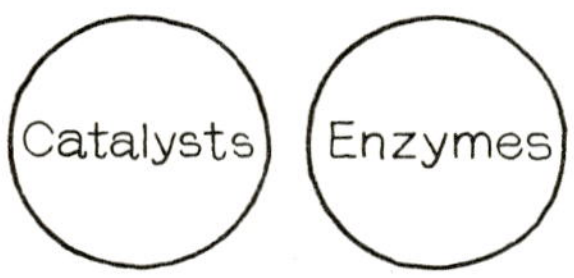

a. ____Catalysts and enzymes are mutually exclusive categories.

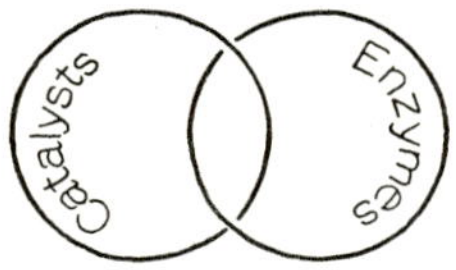

b. ____Some, but not all, enzymes are catalysts.

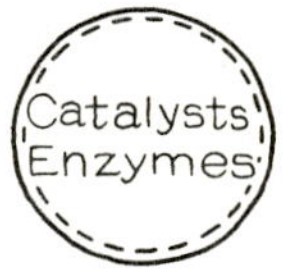

c. ____Catalysts and enzymes are the same thing.

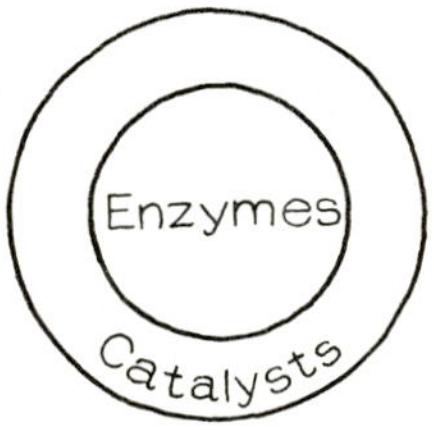

d. ____An enzyme is a kind of catalyst.

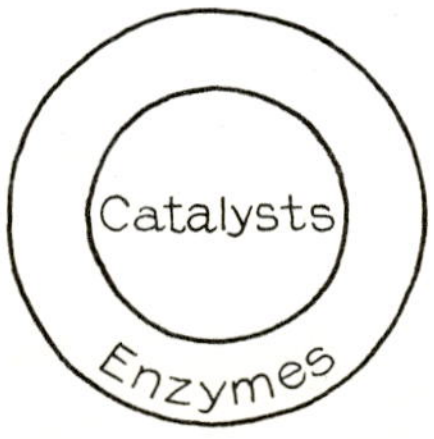

e. ____A catalyst is a kind of enzyme.

3.36
a, b, and c

3.37
no (because enzymes can only accelerate reactions that are spontaneous but slow)

3.39

If you were going to build an enzyme, what features should you include in the active site?

3.38
d

3.40

Knowing that enzymes are proteins, a pair of researchers purified a sample of an enzyme until its composition was 100 percent protein. When they mixed the pure protein with the enzyme's substrate, they found that the protein did not accelerate the reaction. Why did the enzyme lose its activity when it was purified to 100 percent protein?

3.39
shape complementary to substrate(s); binding groups and catalytic groups oriented properly

3.41

a. What is meant by the statement, "Enzymes have a high degree of specificity"?

b. What is the cause of this specificity?

3.40
The enzyme probably contained cocatalysts (metal ions or small organic molecules) that played an important role in catalysis, and the researchers removed them during their purification procedure.

3.42

What is the principal difference between enzymes and ordinary catalysts that makes the former more efficient at accelerating reactions?

3.41

a. A given enzyme catalyzes only one reaction or a few closely related reactions.
b. The close fit between the active site of the enzyme and the substrate.

3.43

What are two methods, in addition to the action of oriented catalytic groups, by which some enzymes are believed to accelerate reactions?

3.42
In an enzyme, catalytic groups are all present and ideally oriented in the active site.

● Go to Unit Four.

help frames for unit three

3.44
3.11

Why the transition state is the substance of highest free energy In frame 3.05 the transition state was defined as the configuration of atoms that has the maximum potential energy during a reaction. The conversion of kinetic energy to potential energy decreases the disorder of energy distribution in a system at cellular temperatures, so the transition state represents the minimum disorder of energy distribution.

In addition, the transition state usually corresponds to the minimum positional disorder. For example:

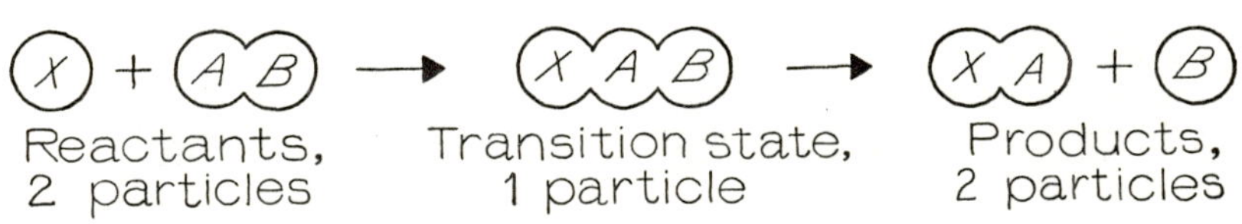

Since the transition state represents the minimum total disorder, it must represent the maximum free energy. (By definition, a decrease in disorder is an increase in free energy; see frame 1.36 if you wish to review this definition.)

● Return to frame 3.11.

3.43
They provide an alternative and faster pathway for the reaction, and they weaken bonds in the substrate by changing its shape.

3.45
3.16

The higher the temperature of a collection of molecules, the faster those molecules are moving. The following diagrams

3.44
no answer

show two simple systems. The events in system *B* are the same as those in system *A*, except that they are happening only half as fast.

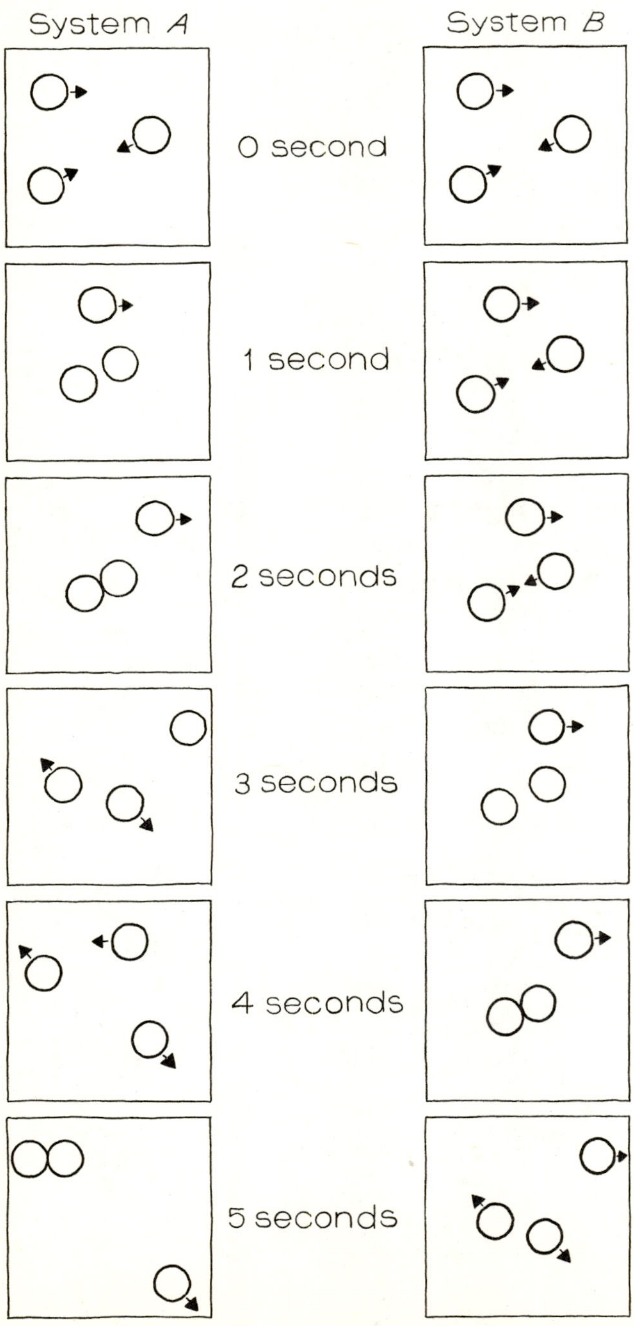

Results:
2 collisions between particles in 5 seconds

Results:
1 collision in 5 seconds

● Now try to answer the question in frame 3.16 again.

3.46
3.24

The answer to the question in frame 3.24 is explained by the fact that catalysts are not consumed or permanently changed by the reactions they catalyze. They are always regenerated by the time the reaction is completed. For example:

3.45
no answer

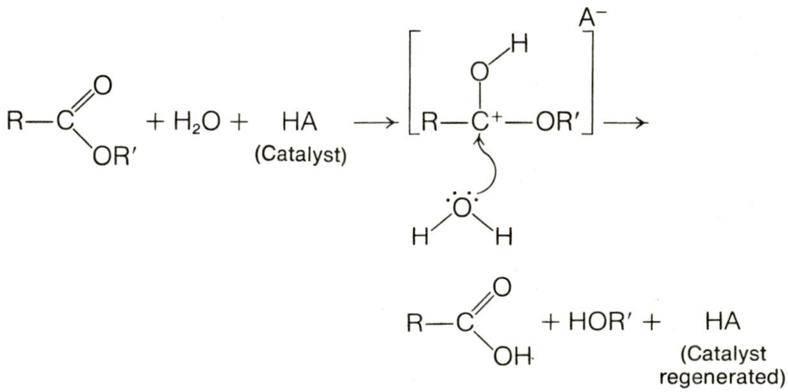

● Now try to answer the question in frame 3.24.

3.47
3.26

As you learned in frame 1.44, the ΔG for a reaction is not affected by the rate at which the reaction takes place. Catalysts influence only the rate of a reaction; they do not change the ΔG, as the following diagrams illustrate:

3.46
no answer

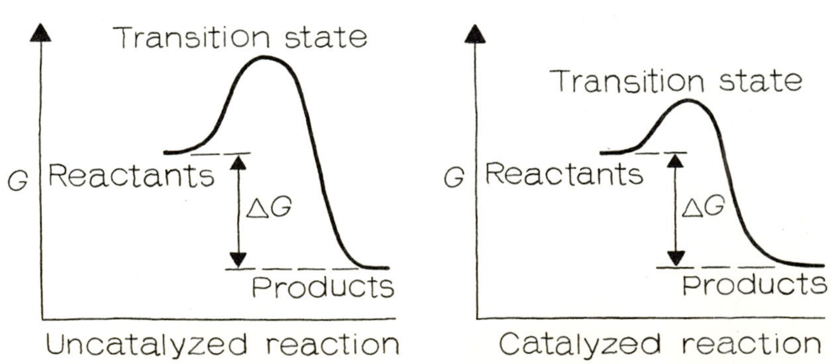

● Return to frame 3.26.

3.48
3.31

Why formation of the enzyme-substrate complex is spontaneous Disorder of position decreases when one or two substrate molecules and an enzyme molecule join to form one enzyme-substrate complex. However, this decrease in positional disorder is more than made up for by the increased disorder of energy distribution that results when bonds are formed between enzyme and substrates.

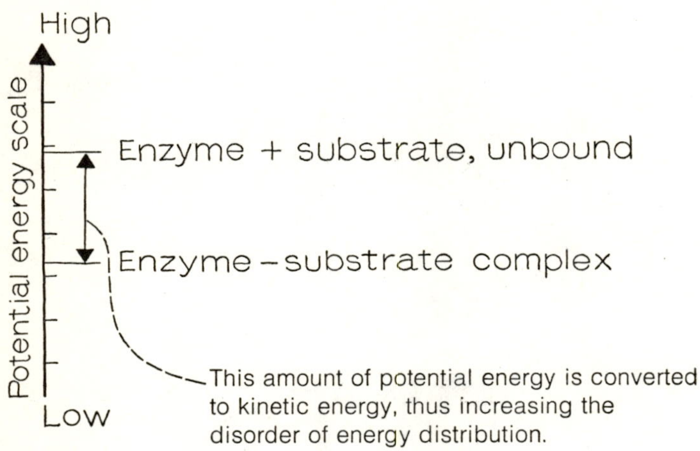

● Return to frame 3.31.

3.47
no answer

3.48
no answer

unit four

reactions in which electrons are transferred

one
introduction

4.01

Some chemical reactions consist of the transfer of electrons from one substance to another. A simple example of such a reaction is the transfer of two electrons from an atom of iron to an atom of sulfur:

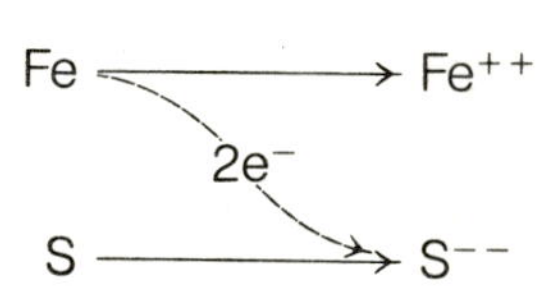

This transfer of electrons results in the formation of a new compound, ferrous sulfide (FeS):

Fe	+	S	$\longrightarrow$	$Fe^{++}S^{--}$
iron metal (grey, melting point = 1540° C)		atomic sulfur (yellow, melting point = 115° C)		ferrous sulfide (black, melting point = 1200° C)

An important point about reactions in which electrons are transferred is that atoms vary in their readiness to accept or give up electrons. Thus, depending on the relative tendencies of the reactants to donate or accept electrons, some electron transfers occur spontaneously and others do not. For example, the transfer of electrons from a strong donor to a weak donor is spontaneous, whereas the transfer of electrons from a weak donor to a strong donor is not. (Chemists have methods of measuring the relative tendency of a substance to donate electrons. Thus we can predict whether a given electron transfer will be spontaneous.)

Like any other exergonic reactions, spontaneous electron transfers can be coupled to endergonic reactions. Spontaneous electron transfers that are coupled to endergonic reactions play a very important role in the breakdown of food in cells that use oxygen. Part of the respiration pathway is a series of electron transfers in which the reactants are arranged in descending order of ability to donate electrons. (Thus the transfers are spontaneous.) By means of intermediates, one of which is ATP, these spontaneous electron transfer reactions are coupled to the endergonic reactions on which such vital processes as growth and movement are based.

Before electron transfers in respiration can be discussed further, some terminology and conventions need to be introduced or reviewed.

4.01
no answer

two

terminology, conventions, and notation

➡ If you are already familiar with the definitions of redox reaction, oxidation, reduction, and half reaction, and if you can balance reactions with respect to charge, go on to frame 4.08. If you miss any questions there, come back to the appropriate frames in this section and review before going on.

4.02 ____

Oxidation and reduction

- Reactions in which electrons are transferred from one substance to another substance are called oxidation-reduction reactions, or redox reactions.

- Loss of electrons is called oxidation. Gain of electrons is called reduction.

If you have trouble remembering the definitions of oxidation and reduction, it is not hard to obtain the definition of oxidation with the following reasoning; the definition of reduction follows because it is the opposite of the definition of oxidation.

1. *Oxidation means adding oxygen.* Although the term "oxidation" now has a wider meaning, it originally referred to the process of combining some substance with the element oxygen.
2. *The oxygen atom readily accepts electrons.* To remember this fact, recall that oxygen in compounds almost always has a valance of −2, indicating that it usually accepts two electrons.
3. *Oxidation of X involves loss of electrons from X.* When an oxygen atom combines with some other substance, which we shall call *X*, the oxygen accepts electrons from *X*. Therefore, oxidation equals loss of electrons from *X*.

4.03 ______________________________
4.11
4.13

Half reactions Free electrons, those not associated with any nucleus, are tremendously unstable. Under ordinary conditions electrons cannot exist free; they can only be passed directly from one substance to another substance. It follows that if a substance loses electrons, some other substance must simultaneously accept those electrons.

4.02
no answer

● Every oxidation is accompanied by a simultaneous reduction, and every reduction is accompanied by a simultaneous oxidation.

Sometimes, for purposes of convenience, reactions are written as if isolated electrons could exist free in the medium. For example:

$$\text{Fe (metal)} \longrightarrow Fe^{++} + 2\ e^-$$

A reaction like this is called a half reaction, because it does not tell what substance accepts the electrons.

This particular half reaction shows
_____an oxidation.
_____a reduction.

Similarly, a half reaction can show electrons being accepted without indicating what substance donates the electrons. For example:

$$Zn^{++} + 2\ e^- \longrightarrow Zn\ \text{(metal)}$$

4.04 ______________________________

Practice in identifying oxidations and reductions Decide whether each of the following half reactions is an oxidation or a reduction and write your answer in the blank.

a. $S + 2\ e^- \longrightarrow S^{--}$

____

b. $Fe^{+++} + e^- \longrightarrow Fe^{++}$

____

c. $Na \longrightarrow Na^+ + e^-$

____

4.03
an oxidation (because Fe metal loses electrons, and oxidation = loss of electrons)

4.05 ______________________________
4.14

Balancing redox reactions with respect to charge Electrons and protons are particles of matter and, like all matter, they cannot be created or destroyed (except in thermonuclear reactions). This implies that the charge resulting from the presence of electrons or protons is also conserved.

4.04
a. reduction
b. reduction
c. oxidation

- In a redox reaction or half reaction, the total charge on reactants equals the total charge on products.

In which of the following reactions is the charge not conserved?
_____first
_____second

$$\tfrac{1}{2}\ Cl_2 + e^- \longrightarrow Cl^-$$

$$Na \longrightarrow Na + e^-$$

4.06 ______________________________

The first reaction in frame 4.05 would be considered balanced because the charge on the left side of the equation equals the charge on the right side. Unbalanced reactions such as the second reaction in frame 4.05 cannot occur in nature.

4.05
second

Is the following reaction balanced?

_____yes

_____no

If not, correct it so that it is.

$$2\ e^- + 2\ H^+ \longrightarrow H_2$$

4.07 ____

Is the following half reaction balanced?

_____yes

_____no

If not, correct it so that it is.

$$Fe^{+++} \longrightarrow Fe^{++} + e^-$$

____

4.06

yes, the reaction is balanced

review questions

4.08 ____

4.02

State the definition of redox reaction.

4.09 ____

4.07

no; the balanced equation would be $Fe^{+++} + e^- \longrightarrow Fe^{++}$ or $Fe \longrightarrow Fe^{++} + 2\ e^-$ (Other answers are possible, as long as total charge on reactants = total charge on products.)

Is the following reaction a redox reaction?

_____yes

_____no

$$H^+Cl^- + Na^+OH^- \longrightarrow Na^+Cl^- + H^+OH^-$$

4.08

a reaction in which electrons are transferred from one substance to another

4.10 ____

Is the following reaction a redox reaction?

_____yes

_____no

$$Fe\ (metal) + Zn^{++}SO_4^{--} \longrightarrow Zn\ (metal) + Fe^{++}SO_4^{--}$$

4.09

no (The charge on each ion does not change, indicating that no electrons are being transferred from one ion to another.)

4.11 ____

Which of the following is (are) half reactions? __________

1. $\frac{1}{2} H_2 + \frac{1}{2} I_2 \longrightarrow HI$
2. $Fe^{++} \longrightarrow Fe^{+++} + e^-$
3. $H^+Cl^- + Ag^+NO_3^- \longrightarrow Ag^+Cl^-$ (precipitate) $+ H^+NO_3^-$
4. $\frac{1}{2} Cl_2 + e^- \longrightarrow Cl^-$

4.12

The following half reaction is
_____a reduction.
_____an oxidation.

$$2\ Cu^{++} + 2\ e^- \longrightarrow Cu_2^{++}$$

4.13

Can an isolated half reaction occur in nature?
_____yes
_____no

4.14

Is the following half reaction balanced?
_____yes
_____no
If not, correct it so that it is.

$$Co^{++} + e^- \longrightarrow Co^{+++}$$

summary

4.15

Oxidation-reduction (redox) reactions are reactions in which electrons are transferred from one substance to another. Every oxidation (loss of electrons) is accompanied by a simultaneous reduction (gain of electrons). For purposes of discussion, however, we may consider half reactions, which consist of only an oxidation or only a reduction. Because charge is conserved, redox reactions and half reactions should be written in balanced form; that is, charge on reactants should equal charge on products.

4.10
yes (The charge on Fe goes from 0 to +2 and the charge on Zn goes from +2 to 0. Iron metal transferred 2 electrons to the Zn^{++} ion.)

4.11
2 and 4

See frame 4.03 if you missed this question.

4.12
a reduction (because it is a gain of electrons)

4.13
no

See frame 4.03 if you missed this question.

4.14
no

$Co^{+++} + e^- \longrightarrow Co^{++}$,
$Co^{++} + 2\ e^- \longrightarrow Co$,
$Co^{++} + e^- \longrightarrow Co^+$, and so on.

See frame 4.05 if you missed this question.

4.15
no answer

three

types of redox reactions in living organisms

The redox reactions that occur during respiration can be transfers of electrons to and from metal ions, or they can be transfers of hydrogen atoms from one molecule to another. Examples of both types of reaction will be discussed here.

4.16 ______

Electron transfers to and from metal ions The cytochromes are a group of compounds that contain metal ions and that take part in redox reactions in the cell.

➡ If you do not know what cytochromes are, go to frame 4.17. If you do, test your knowledge by answering the questions in frame 4.18. If you do not answer the questions there correctly, read through frame 4.17 to refresh your memory.

4.17 ______

Cytochromes are found in certain membranes of cells that utilize oxygen. Every cytochrome consists of two major com-

4.16
no answer

ponents: a protein and an active group, which contains nitrogen atoms and iron, in addition to carbon, hydrogen, and oxygen. The active group is held in place by covalent bonds between the active group and the protein.
The following sketch of a cytochrome and its components has been simplified for purposes of illustration. The three-dimensional structure of the protein of cytochromes is actually more complicated.

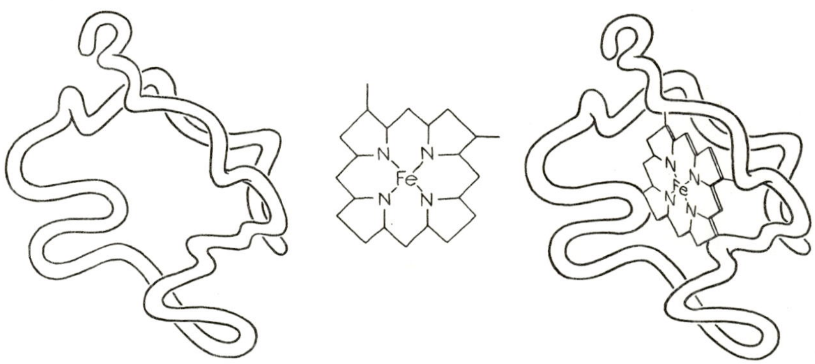

Protein   Active group   Complete cytochrome

The iron in a cytochrome is present as Fe^{++} or Fe^{+++}. The iron can be converted from one state to the other by loss or gain of an electron. For example:

$$\text{cytochrome-Fe}^{++} \rightleftarrows \text{cytochrome-Fe}^{+++} + e^-$$

There are three major types of cytochrome in eucaryotic cells: cytochrome *a*, cytochrome *b*, and cytochrome *c*, plus many variants. This discussion will be restricted to these three types. They differ slightly in the structure of the iron-containing active group and in the way the active group is attached to the protein.

4.18 ______________________________

4.16

a. Cytochromes contain an ion of what metal? __________

b. In cytochromes this ion can have a charge of ____ or ____.

c. The major part of a cytochrome molecule is
____carbohydrate.
____lipid.
____protein.

4.17
no answer

d. Which part of the cytochrome molecule participates in redox reactions? ________.

4.19 ________

There are two types of electron transfers that involve cytochromes:

1. Electrons can be transferred from one kind of cytochrome molecule to another. For example:

$$\text{cytochrome } c\text{-Fe}^{++} + \text{cytochrome } b\text{-Fe}^{+++} \longrightarrow \text{cytochrome } c\text{-Fe}^{+++} + \underline{\qquad\qquad}$$

(fill in the blank)

2. Electrons can also be transferred between cytochromes and other molecules. For example:

$$2 \text{ cytochrome } a\text{-Fe}^{++} + \tfrac{1}{2}\,O_2 \longrightarrow 2 \text{ cytochrome } a\text{-Fe}^{+++} + O^{--*}$$

*O^{--}, which is produced by this reaction, is so unstable that it does not exist for any length of time. In the cell it combines immediately with 2 H^+ to form H_2O.

4.18
a. iron
b. +2 or +3
c. protein
d. the iron ion

4.20 ________

In the second reaction in the preceding frame, two electrons are transferred from ________ to ________.

4.19
cytochrome *b*-Fe^{++}

4.21 ________

Is the following reaction an oxidation or a reduction? ________

$$\text{cytochrome } c\text{-Fe}^{++} \longrightarrow \text{cytochrome } c\text{-Fe}^{+++} + e^-$$

4.20
cytochrome *a*-Fe^{++} to oxygen ($\tfrac{1}{2}O_2$)

4.22 ________

Redox reactions in which hydrogen atoms are transferred
Electron transfer may consist of transfer of hydrogen atoms. For example, in the following redox reaction, electrons are lost by succinic acid:

4.21
oxidation

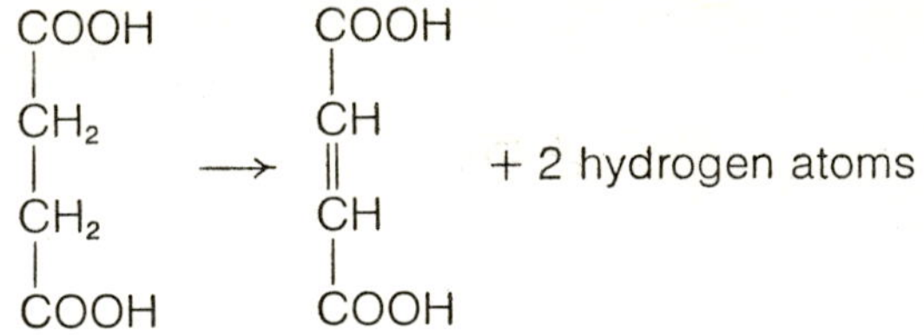

succinic acid fumaric acid

We can see the loss of electrons more clearly if we rewrite the structure for succinic acid so that a dot represents each electron in the bonds between succinic acid and the two hydrogen atoms that leave:

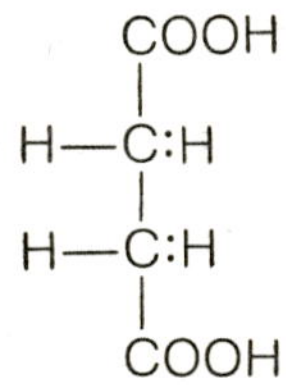

Each bond is a shared pair of electrons, one electron donated by the carbon atom and one electron donated by the hydrogen atom. In such covalent bonds between hydrogen and carbon, the electrons are not shared completely equally. Instead, each electron in the pair spends somewhat more than half its time with the carbon nucleus. Thus the carbon atom bonded to hydrogen has the equivalent of somewhat more than one electron to itself.† When the bond is broken and the hydrogen atom leaves with its electron, the carbon atom has exactly one electron to itself:

$$\mathrm{C{:}H} \longrightarrow \mathrm{C\cdot} + \mathrm{\cdot H}$$

The change from somewhat more than one electron to one electron constitutes loss of electrons.

†This discussion ignores the other three electrons in the outer shell of carbon.

4.23 ______________________________

Hydrogen atoms, which are produced by the reaction in frame 4.22, are so reactive that they do not actually exist as atoms for long, if at all. Hydrogen atoms either attach themselves to another molecule:

4.22
no answer

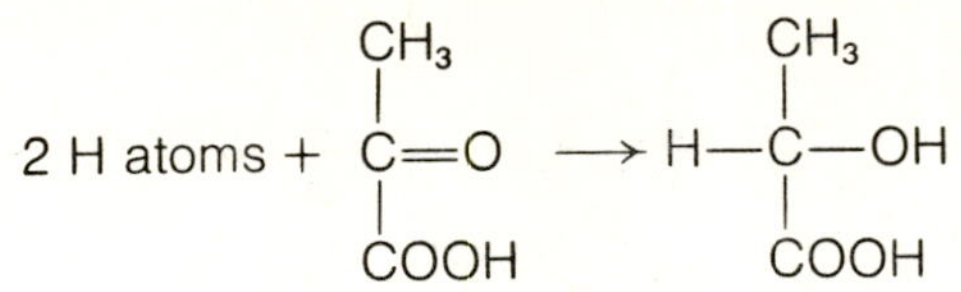

pyruvic acid lactic acid

or they transfer their electrons to another substance, thus becoming H^+ ions or protons:

H atom + cytochrome *c*-Fe^{+++} ⟶ H^+ + cytochrome *c*-Fe^{++}

(In aqueous solution, protons are bound to water molecules and thus exist as H_3O^+.)

● If you do not understand why H^+ is the same as a proton, see frame 4.79 for an explanation.

4.24

To summarize,

4.23
no answer

● Biological redox reactions can consist of either of the following:

1. Transfer of electrons (for example, from one cytochrome to another).
2. Transfer of hydrogen atoms (electrons plus protons) from one substance to another.

4.25

Definitions of "oxidized" and "reduced"

4.24
no answer

● Oxidation is loss of electrons; when a compound loses electrons it becomes oxidized.

Reduction is gain of electrons; when a compound gains electrons it becomes reduced.

a. The electric charge on an oxidized cytochrome will be ____.

b. The electric charge on a reduced cytochrome will be ____.

4.26

Which of the following reactions is(are) oxidation(s)?

____

4.25
a. +3
b. +2

1. $Zn^{++} + 2\ e^- \longrightarrow Zn$ (metal)

2.
$$\underset{\text{ethanol}}{CH_3{-}\overset{\displaystyle H}{\underset{\displaystyle H}{\overset{|}{\underset{|}{C}}}}{-}OH} \longrightarrow \underset{\text{acetaldehyde}}{CH_3{-}\overset{\displaystyle H}{\overset{|}{C}}{=}O} + 2 \text{ hydrogen atoms}$$

3. Na (metal) $\longrightarrow Na^+ + e^-$

4. cytochrome *b*-$Fe^{++} \longrightarrow$ cytochrome *b*-$Fe^{+++} + e^-$

4.27 ______

In the following half reaction, lactic acid becomes
____reduced.
____oxidized.

lactic acid $\longrightarrow$ pyruvic acid + 2 H atoms

4.26
2,3,and 4

4.28 ______

Write the half reaction that shows what happens when cytochrome *b*-Fe^{+++} becomes reduced.

4.27
oxidized

4.29 ______

FAD and NAD$^+$, important biological compounds that participate in redox reactions Now that you have the ideas of oxidation and reduction well in your grasp, it is a good time to introduce two more compounds that participate in redox reactions in the cell. Since both of these compounds have long names, they will be referred to here by their abbreviations, FAD and NAD$^+$.

- It is not necessary to know the full names and structures of FAD and NAD$^+$ to complete this program, but if you are curious, see frame 4.82.

FAD accepts two hydrogen atoms at a time:

$$FAD + 2\ H^+ + 2\ e^- \longrightarrow FADH_2$$

4.28
cytochrome *b*-$Fe^{+++} + e^- \longrightarrow$ cytochrome *b*-Fe^{++}

(By convention, a hydrogen atom is written as $H^+ + e^-$ in redox reactions.)

a. The preceding half reaction is
_____an oxidation.
_____a reduction.

Optional section for those who are interested.

The part of the FAD molecule that is involved in the redox reaction looks like this:

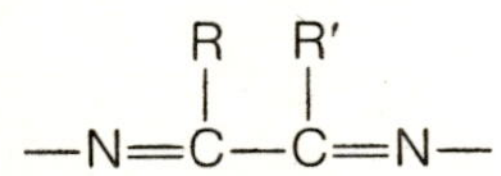

When FAD is reduced to $FADH_2$, one hydrogen atom is added at each of the two nitrogen atoms shown. Note the change that has occurred in the number and position of the double bonds:

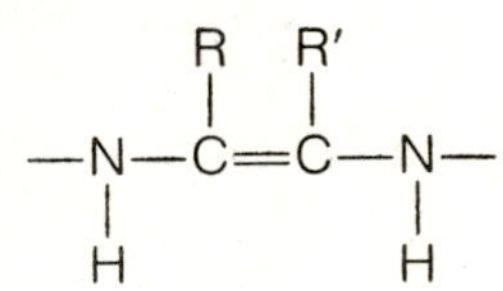

NAD^+ accepts two electrons and one proton:

$$NAD^+ + 2\,H^+ + 2\,e^- \longrightarrow NADH + H^+$$

Note that there is a positive charge on NAD^+.

b. Is the preceding reaction balanced with respect to charge?
_____yes
_____no

Optional section for those who are interested.

The part of the NAD^+ molecule involved in the redox reaction looks like this (note the positive charge on the nitrogen atom):

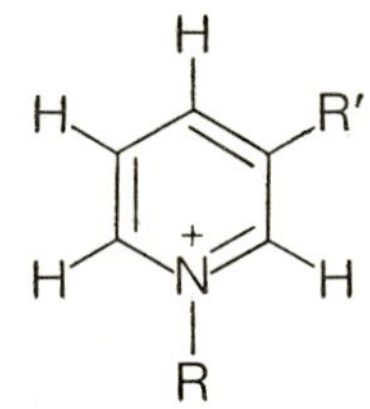

The two electrons and one proton join the molecule at the top of the ring, causing a change in the number and position of double bonds. As a result, the nitrogen atom on NADH is no longer positively charged:

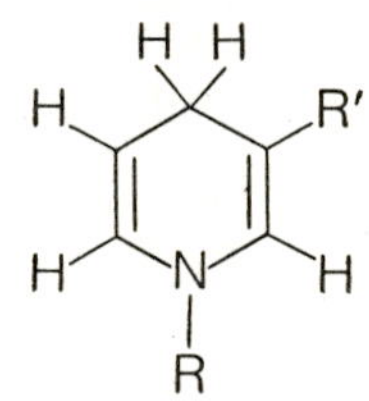

4.30

Complete this equation:

$$FADH_2 + NAD^+ \longrightarrow \underline{\qquad\qquad\qquad}$$

4.29

a. a reduction

b. yes

4.31

There are two mistakes in the following equation. What are they?

$$NAD + 2\ H^+ + 2\ e^- \longrightarrow NADH_2$$

4.30

$FAD + NADH + H^+$

4.32 ______________________________

a. The following half reaction is
_____an oxidation.
_____a reduction.

$$\text{lactic acid} \longrightarrow \text{pyruvic acid} + 2\ H^+ + 2\ e^-$$

b. The following half reaction is
_____an oxidation.
_____a reduction.

$$NAD^+ + 2\ e^- + 2\ H^+ \longrightarrow NADH + H^+$$

➡ If you feel that you have had enough practice in writing equations that involve NAD^+, FAD, and cytochromes, you may skip to frame 4.35.

4.31
NAD should be NAD^+; $NADH_2$ should be $NADH + H^+$.

4.33 ______________________________

Combine the two half reactions in frame 4.32 into a redox reaction, crossing out any substances that appear on both sides of the arrow.

4.32
a. an oxidation
b. a reduction

4.34 ______________________________

Write the following redox reaction as two half reactions. Then identify the half reaction that is a reduction.

$$FADH_2 + 2\ \text{cytochrome}\ b\text{-}Fe^{+++} \longrightarrow FAD + 2\ \text{cytochrome}\ b\text{-}Fe^{++} + 2\ H^+$$

4.33

$$\text{lactic acid} + NAD^+ \longrightarrow \text{pyruvic acid} + NADH + H^+$$

4.35 ______________________________
4.32

Notation for biological redox reactions There is a system of notation for writing redox reactions that you will probably encounter in biology texts. In this system of notation a reaction like

$$A^{++} + B \longrightarrow A + B^{++}$$

4.34

$$FADH_2 \longrightarrow FAD + 2\ H^+ + 2\ e^-$$

$$2\ \text{cytochrome}\ b\text{-}Fe^{+++} + 2\ e^- \longrightarrow 2\ \text{cytochrome}\ b\text{-}Fe^{++}$$

The second half reaction shown here is a reduction.

is written:

In this particular reaction, electrons are transferred from _____ to _____.

4.36 ______________________________

Draw a dotted line to show the path of electrons in the redox reaction represented as:

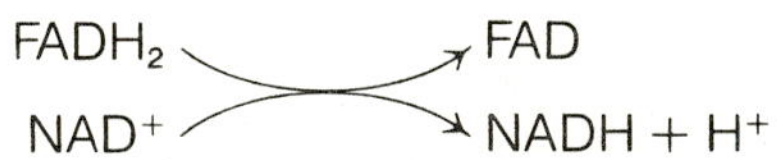

4.35
B to *A*

If you need help, go to frame 4.80.

4.37 ______________________________

a. Write the following redox reaction using the new notation.

b. Then trace the path of the electrons with a dotted line.

$$\text{succinic acid} + \text{FAD} \longrightarrow \text{fumaric acid} + \text{FADH}_2$$

4.36

4.38 ______________________________

Definitions of "oxidizing agent" and "reducing agent"

If you can correctly answer the following two questions, skip to frame 4.42.

a. An oxidizing agent
_____accepts electrons.
_____donates electrons.

b. The reducing agent in the following reaction is ______________________.

$$\text{FADH}_2 + 2\ \text{cytochrome } b\text{-Fe}^{+++} \longrightarrow$$
$$\text{FAD} + 2\ \text{H}^+ + 2\ \text{cytochrome } b\text{-Fe}^{++}$$

4.37

a. 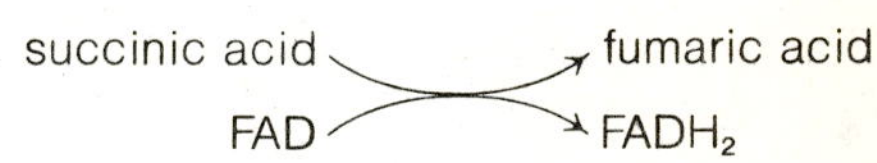

b. 

Consider the following redox reaction:

$$\text{FAD} + \text{succinic acid} \longrightarrow \text{FADH}_2 + \text{fumaric acid}$$

In this reaction, FAD accepts the electrons and protons from succinic acid, and so we could say that FAD is responsible for oxidizing succinic acid. In other words, FAD acts as an oxidizing agent. Similarly, succinic acid donates the electrons that reduce FAD to $FADH_2$. In this reaction, succinic acid is a reducing agent.

● c. An oxidizing agent oxidizes other substances; hence it is an electron
_____acceptor.
_____donor.

d. A reducing agent reduces other substances; hence it is an electron
_____acceptor.
_____donor.

4.39 ______________________________

The reducing agent in the following reaction is __________.

$$\text{NADH} + \text{H}^+ + \text{FAD} \longrightarrow \text{NAD}^+ + \text{FADH}_2$$

4.38
a. accepts electrons
b. $FADH_2$
c. acceptor
d. donor

4.40 ______________________________

The oxidizing agent in the following reaction is
__________________.

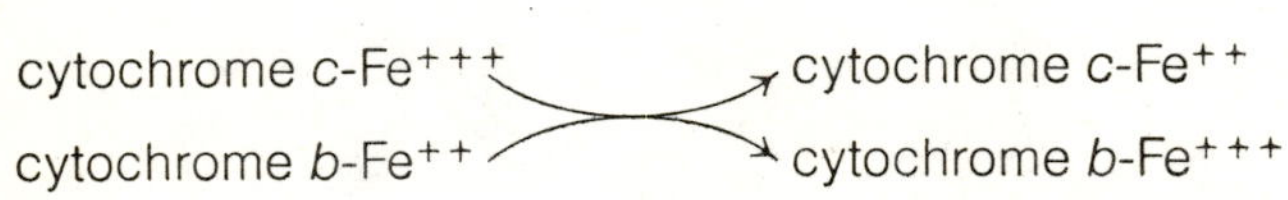

4.39
NADH

____

summary

4.41 ______________________________

Redox reactions can consist of transfers of electrons or transfers of hydrogen atoms (electrons plus protons). Compounds that participate in redox reactions in the cell include the cytochromes, NAD^+, and FAD. Biological redox reactions are often written in a curved-arrows notation. An oxidizing agent accepts electrons, thus oxidizing another substance; a reducing agent donates electrons, thus reducing another substance.

4.40
cytochrome *c*-Fe^{+++}

4.41
no answer

four

recognizing the redox reactions in glycolysis and the krebs cycle

4.42

4.38

➡ If you are already familiar with glycolysis and the Krebs cycle, you may skip to the gray arrow at the end of this frame.

The cellular breakdown of any food molecule occurs not in a single chemical reaction but rather in a sequence of many reactions. For example, in cells that use oxygen, the breakdown of glucose to carbon dioxide occurs via a pathway of nineteen different reactions. In the first half of the pathway, one molecule of glucose is converted to two molecules of pyruvic acid. This portion of the pathway is called glycolysis. The pyruvic acid then undergoes a series of reactions known as the Krebs cycle.‡ Other types of food molecules, such as amino acids and

‡The Krebs cycle is also called the citric acid cycle or the tricarboxylic acid cycle.

fatty acids, can also enter the Krebs cycle after being converted to a reactant of one of the steps of the Krebs cycle. Some of the reactions that occur in glycolysis and the Krebs cycle are redox reactions.

➡ If you would like to apply what you have learned about redox reactions to glycolysis and the Krebs cycle, go to frame 4.43. If you do not want more practice, skip to frame 4.46.

4.43

Examine the charts of glycolysis and the Krebs cycle that precede the Glossary at the end of the book. In the space below copy the numbers of any redox reactions you find there.

Glycolysis	Krebs cycle

4.42
no answer

4.44

The following chart shows the redox reactions that occur in glycolysis and the Krebs cycle. In all these reactions, protons are transferred along with the electrons.

4.43
There is one redox reaction in glycolysis (6); there are five redox reactions in the Krebs cycle (11, 14, 15, 17, 19).

GLYCOLYSIS

reaction 6

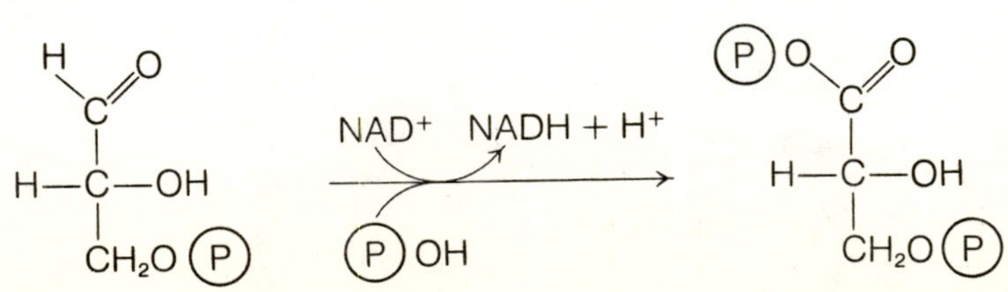

glyceraldehyde-3-phosphate

1,3-diphosphoglyceric acid

KREBS CYCLE

reaction 11

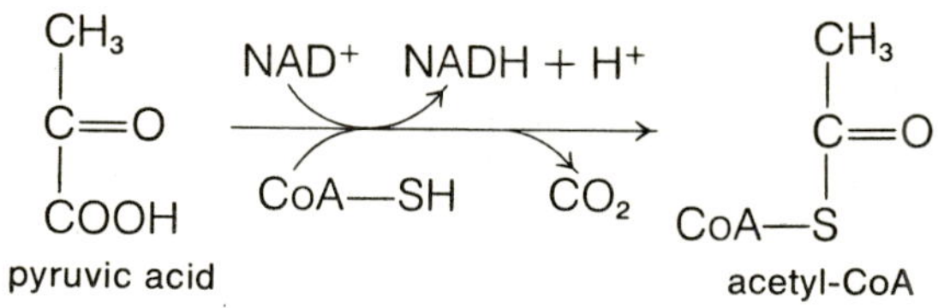

reaction 14

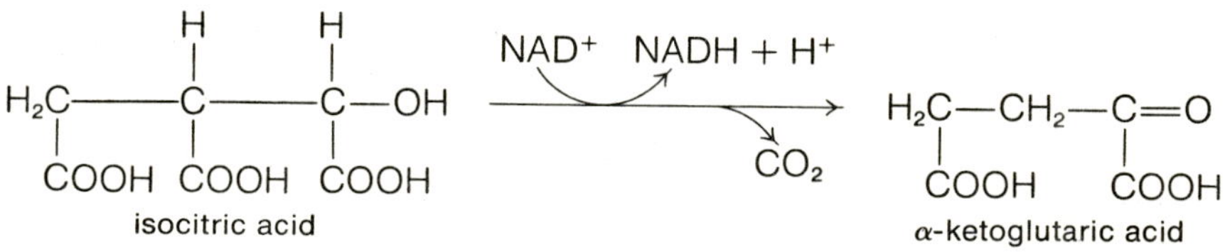

reaction 15

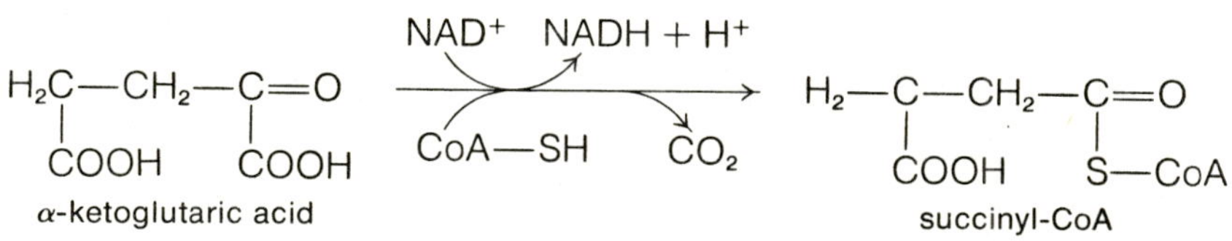

reaction 17

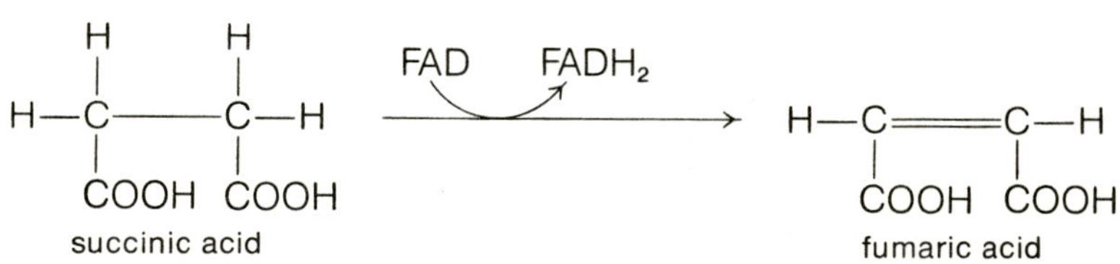

reaction 19

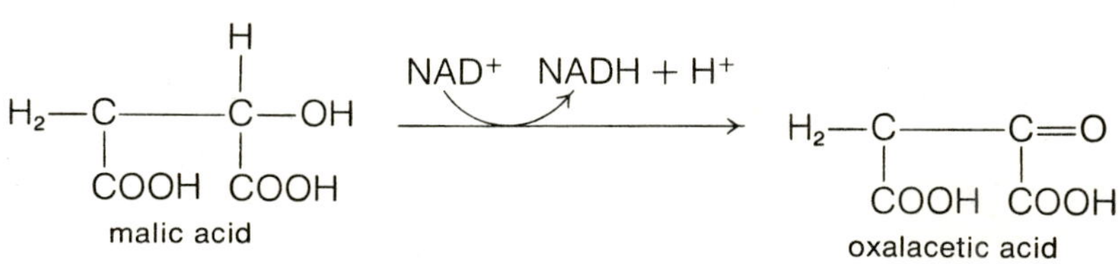

a. In every redox reaction occurring in glycolysis and the Krebs cycle, derivatives of the glucose molecule

____gain electrons.

____lose electrons.

b. In other words, these derivatives are
____oxidized.
____reduced.

4.45 ______________________

What is the oxidizing agent in each of the redox reactions in the chart in frame 4.44?

Reaction number	Oxidizing agent
6	________
11	________
14	________
15	________
17	________
19	________

4.44
a. lose electrons
b. oxidized

4.45
6: NAD^+
11: NAD^+
14: NAD^+
15: NAD^+
17: FAD
19: NAD^+

five

measuring the ΔG of redox reactions

4.46 ______________________________

4.42

Substances differ in their ability to accept or donate electrons. For example, FAD tends to attract electrons more strongly than NAD^+ does. Therefore, which of the following reactions will proceed spontaneously?

_____reaction 1

_____reaction 2

1. $FADH_2 + NAD^+ \longrightarrow FAD + NADH + H^+$

2. $FAD + NADH + H^+ \longrightarrow FADH_2 + NAD^+$

4.47 ______________________________

As the preceding frame demonstrates, the strength with which each of the reactants and products attracts electrons determines whether a reaction is spontaneous. Thus ΔG

4.46
reaction 2

depends on the readiness with which the reactants and products gain or lose electrons.

4.48

How the ability of a substance to accept electrons is measured

4.47
no answer

➡ If you are already familiar with the method by which reduction potentials are measured, you may skip to frame 4.49 if you wish.

Read the following material for understanding. A summary of the important points to remember appears in frame 4.49.

Suppose that we want to measure the tendency of cytochrome *a*-Fe^{+++} to accept electrons. The reduction of cytochrome *a*-Fe^{+++} can be written as follows:

$$\text{cytochrome } a\text{-Fe}^{+++} + e^- \longrightarrow \text{cytochrome } a\text{-Fe}^{++}$$

We prepare a mixture containing one mole of the oxidized form (cytochrome *a*-Fe^{+++}) and one mole of the reduced form (cytochrome *a*-Fe^{++}). Initially, the concentrations of the two forms are equal, but the concentrations will change as the reaction moves toward equilibrium.

If cytochrome *a*-Fe^{+++} has strong electron affinity, the equilibrium will lie to the right:

$$\text{cytochrome } a\text{-Fe}^{+++} + e^- \rightleftharpoons \text{cytochrome } a\text{-Fe}^{++}$$

The reaction will therefore use up electrons. We can supply electrons by putting a strip of metal, such as platinum, into the solution. The metal atoms donate electrons, and as a result the strip of metal becomes positively charged:

$$\text{metal atoms} \longrightarrow \text{metal ions (+charged)} + \text{electrons (−charged)}$$

On the other hand, if cytochrome *a*-Fe^{+++} does not have strong electron affinity, the equilibrium will lie to the left:

$$\text{cytochrome } a\text{-Fe}^{+++} + e^- \rightleftharpoons \text{cytochrome } a\text{-Fe}^{++}$$

Electrons will be released by this reaction and will build up a negative charge on the strip of platinum. This charge can be considered electron pressure.

To find out whether the strip of platinum is charged positively or negatively and thus to measure the strength of cytochrome *a*'s electron affinity or electron pressure, we need a standard to which we can compare cytochrome *a*. A widely accepted standard is the following reaction:

$$H^+ + e^- \rightleftharpoons \frac{1}{2} H_2$$

To use this reaction as a standard, we place a strip of platinum in a solution of H^+ ions at one molar concentration. Then H_2 gas at a pressure of one atmosphere is bubbled through the solution so that the concentrations of the oxidized form (H^+) and the reduced form (H_2) are equal. Such an arrangement is called a half cell. The mixture of one molar cytochrome *a*-Fe^{++} and one molar cytochrome *a*-Fe^{+++} is also called a half cell.

If we connect the platinum strip of the hydrogen half cell by a wire to the platinum strip of the cytochrome *a* half cell and if we complete the circuit by providing a salt bridge§ between the two cells, electrons can pass through the wire. The electrons will naturally flow from a region of higher electron pressure to a region of lower electron pressure. Thus we can compare the electron pressures of the two half cells by noting which way and how strongly the electrons flow.

If the electron pressure of the cytochrome *a* half cell is greater than that of the hydrogen half cell, electrons will flow spontaneously from the cytochrome *a* half cell to the hydrogen half cell. This electron pressure is measured as a voltage.

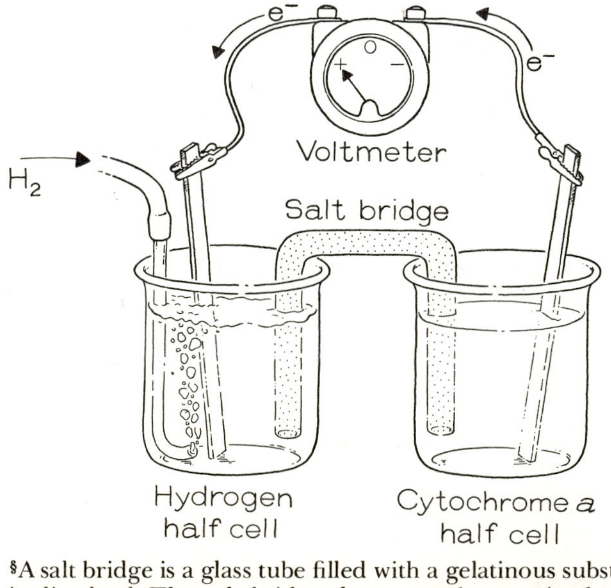

§A salt bridge is a glass tube filled with a gelatinous substance in which a salt is dissolved. The salt bridge does not take part in the chemical reactions occurring in the half cells, but it must be present to complete the electrical circuit.

If, on the other hand, the electron pressure of the cytochrome *a* half cell is less than the electron pressure of the hydrogen half cell, electrons will flow spontaneously in the opposite direction.

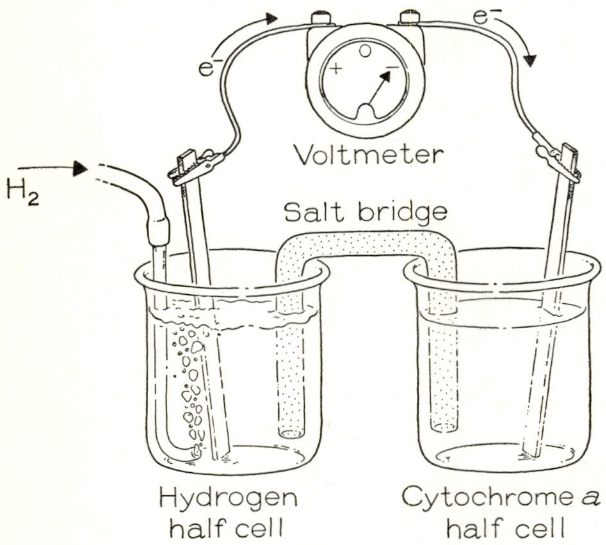

For the sake of convenience, the electron pressure of the hydrogen half cell is defined as zero. Consequently, the absolute value of the voltage measured in the experiment gives the absolute value of cytochrome *a*'s electron pressure or electron affinity. In this book the absolute value of the voltage will be used to express electron affinity. When the voltage is used in this way, it is called the reduction potential, and its sign is determined by the following convention. If cytochrome *a* has a greater affinity for electrons than hydrogen does, the sign of the reduction potential is positive. The sign is negative if cytochrome *a* has a lower electron affinity than hydrogen does.

Half cells can be made up for other compounds, and the electron affinity of each can be compared with the electron affinity of the hydrogen half cell. In this manner, reduction potentials for these other compounds are obtained. The reduction potential at standard conditions is symbolized by E_0. E_0' is the symbol for reduction potential when $p\text{H} = 7$ and all other conditions are standard. Following are the values of E_0' for a number of compounds with which you are familiar:

Compound	E_0' (volts)
NAD^+	−0.320
pyruvic acid	−0.190
FAD	−0.050
cytochrome *a*-Fe^{+++}	+0.290
O_2	+0.816

The more positive the value of E_0', the more strongly the compound attracts electrons.

Which would you expect to have a more positive value of E_0'?

_____a strong reducing agent

_____a strong oxidizing agent

4.49 ________________________________

4.48
4.53

4.48
a strong oxidizing agent

Main points to remember about redox potential A standard half cell is a mixture of oxidized and reduced forms of a substance at one molar concentration and standard temperature, pressure, and *p*H.

The standard reduction potential (E_0) of a substance is obtained by comparing the electron pressure of the standard half cell of that substance with the electron pressure of a standard hydrogen half cell.

E_0' is the symbol for reduction potential when $pH = 7$ and all other conditions are standard.

Reduction potential expresses the readiness with which a substance accepts electrons:

A substance with a very positive reduction potential readily accepts electrons and is a strong oxidizing agent.

A substance with a very negative reduction potential readily donates electrons and is a strong reducing agent.

Electrons flow spontaneously from compounds with less positive reduction potentials to compounds with more positive reduction potentials:

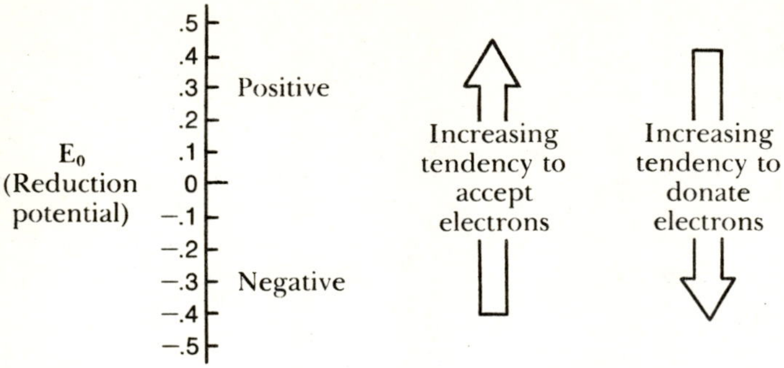

4.50

Consider the following reaction:

$$\text{NADH} + \text{H}^+ + \text{FAD} \longrightarrow \text{NAD}^+ + \text{FADH}_2$$

$$\text{FAD: } E_0' = -0.05 \text{ volts} \qquad \text{NAD}^+\text{: } E_0' = -0.32 \text{ volts}$$

4.49
no answer

a. Which compound tends more strongly to accept electrons?
____NAD^+
____FAD

b. Will this reaction proceed spontaneously?
____yes
____no

c. $\Delta G'$ for the reaction will be
____negative.
____positive.

4.51

Which of the following two reactions has a stronger tendency to proceed spontaneously?
____reaction 1
____reaction 2

4.50
a. FAD
b. yes
c. negative

1.

$$\text{cytochrome } a\text{-Fe}^{++} + \tfrac{1}{2}\,\text{O}_2 + 2\,\text{H}^+ \longrightarrow \text{cytochrome } a\text{-Fe}^{+++} + \text{H}_2\text{O}$$

$$\text{O}_2\text{: } E_0' = +0.816 \text{ volts} \qquad \text{cytochrome } a\text{: } E_0' = +0.290 \text{ volts}$$

2.

$$\text{NADH} + \text{H}^+ + \tfrac{1}{2}\,\text{O}_2 \longrightarrow \text{NAD}^+ + \text{H}_2\text{O}$$

$$\text{O}_2\text{: } E_0' = +0.816 \text{ volts} \qquad \text{NAD}^+\text{: } E_0' = -0.320 \text{ volts}$$

If you need help, go to frame 4.81.

4.52

Relationship between ΔG and redox potential The value of $\Delta G'$ for a redox reaction depends on the reduction potentials of the reactants and products. (This should not be surprising because, as you have seen in the preceding two frames, reduction potentials can be used to predict whether a redox reaction will proceed spontaneously and to predict which of two redox reactions has a greater tendency to proceed spontaneously.)

The relationship between $\Delta G'$ and the reduction potentials can be expressed as follows:

$$\Delta G' = -(\text{number of electrons transferred}) \cdot (23{,}063) \cdot (E_0' \text{ of electron acceptor} - E_0' \text{ of electron donor})$$

$\Delta G'$ will be negative if (E_0' of electron acceptor − E_0' of electron donor) is

_____ negative.
_____ positive.

4.51
reaction 2

4.53

The quantity (E_0' of electron acceptor − E_0' of electron donor) will be positive and $\Delta G'$ will be negative as long as E_0' of electron acceptor is more positive than E_0' of electron donor.

This occurs when electrons are transferred

_____ from compounds with high electron affinity to compounds with low electron affinity.
_____ from compounds with low electron affinity to compounds with high electron affinity.

Review frame 4.49 if you need help.

4.52
positive

4.54

The larger the difference in reduction potential between the electron acceptor and the electron donor,

_____ the larger the absolute value of $\Delta G'$.
_____ the smaller the absolute value of $\Delta G'$.

4.53
from compounds with low electron affinity to compounds with high electron affinity

4.54
the larger the absolute value of $\Delta G'$

4.55

$\Delta G'$ for the following reaction is $-10{,}000$ cal/mole:

cytochrome *b*-Fe^{++} + cytochrome *c*-Fe^{+++} $\longrightarrow$
cytochrome *b*-Fe^{+++} + cytochrome *c*-Fe^{++}

Which compound attracts electrons more strongly?
____cytochrome *c*-Fe^{+++}
____cytochrome *b*-Fe^{+++}

4.55
cytochrome *c*-Fe^{+++}

4.56

$\Delta G'$ for the following reaction is $+3{,}700$ cal/mole:

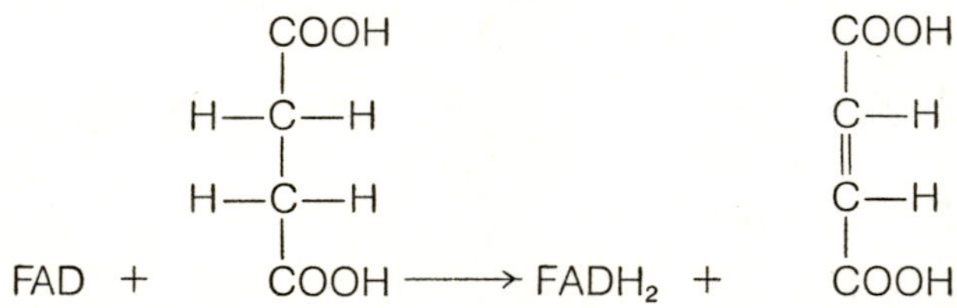

succinic acid   fumaric acid

Which compound attracts electrons more strongly?
____fumaric acid
____FAD

4.56
fumaric acid

summary

4.57

The standard reduction potential (E_0) of a substance is determined by comparing the electron pressure of the substance at standard conditions to the electron pressure of hydrogen at standard conditions. E_0' is the reduction potential measured when pH = 7 and all other conditions are standard. The more positive the E_0 or E_0', the more readily the substance accepts electrons. A redox reaction is spontaneous (has a negative ΔG) if the electrons are being transferred from a compound with a less positive reduction potential to a compound with a more positive reduction potential; in other words, from a weaker acceptor of electrons to a stronger acceptor of electrons.

4.57
no answer

six

the electron transport chain and final review of redox reactions

4.58

➡ If you did the practice exercise in Unit Four, Section Four, "Recognizing the Redox Reactions in Glycolysis and the Krebs Cycle," go to frame 4.59; otherwise, continue with this frame.

Six reactions in glycolysis and the Krebs cycle are redox reactions. An example is the following:

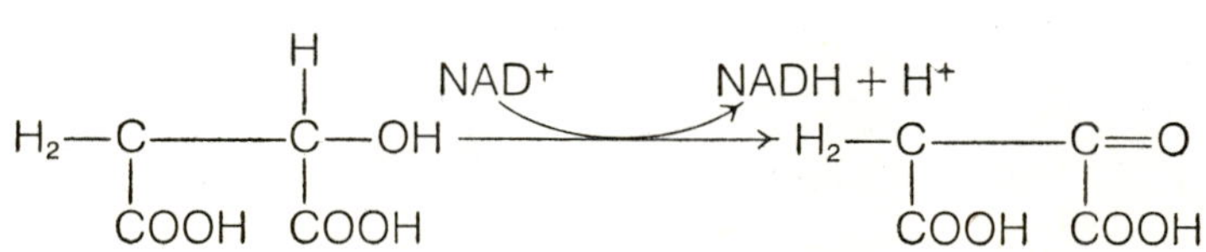

(The molecule on the left was derived, through many changes, from parts of glucose molecules.)

All six of the redox reactions in glycolysis and the Krebs cycle follow the pattern of this example, because in each reaction electrons are removed from derivatives of glucose and transferred to NAD^+ or FAD.

● Go on to frame 4.60.

4.59

As you found by doing the practice exercise in Section Four of this unit, during the oxidation-reduction reactions of glycolysis and the Krebs cycle, electrons are transferred from derivatives of glucose to NAD^+ or FAD.

4.58
no answer

4.60
4.58

For each molecule of glucose that goes through glycolysis and the Krebs cycle,

10 molecules of NAD^+ are reduced to $NADH + H^+$

and

2 molecules of FAD are reduced to $FADH_2$.

4.59
no answer

The subsequent fate of the NADH and $FADH_2$ is very important to the cell. The electrons on $FADH_2$ and NADH are passed on to a series of compounds, each of which is a stronger electron acceptor than the previous compound in the series. The final electron acceptor is oxygen.

Because the electrons are being passed from compounds with less positive reduction potential to compounds with more positive reduction potential, this transfer of electrons is

____spontaneous.
____not spontaneous.

4.61

In this spontaneous series of redox reactions, called the electron transport chain, there are three steps in which $\Delta G'$ is negative enough to drive forward the following reaction:

4.60
spontaneous

$$ADP + H_3PO_4 \longrightarrow ATP + H_2O \qquad (\Delta G' = +8{,}000 \text{ cal/mole})$$

(Exactly how the cell uses the free energy released by these three redox reactions to drive forward the synthesis of ATP is not yet known. However, this question has been and is the subject of much research, and a number of different models have been proposed to explain the mechanism of coupling between the redox reactions and ATP production.)

The ATP synthesized during electron transport can subsequently be converted back to ADP and H_3PO_4. This conversion is an exergonic reaction, and it can be coupled to many of the endergonic reactions on which such important cell activities as growth and movement depend.

The following diagram illustrates how ATP acts as an intermediary in coupling the spontaneous redox reactions of the electron transport chain to vital endergonic reactions of the cell.

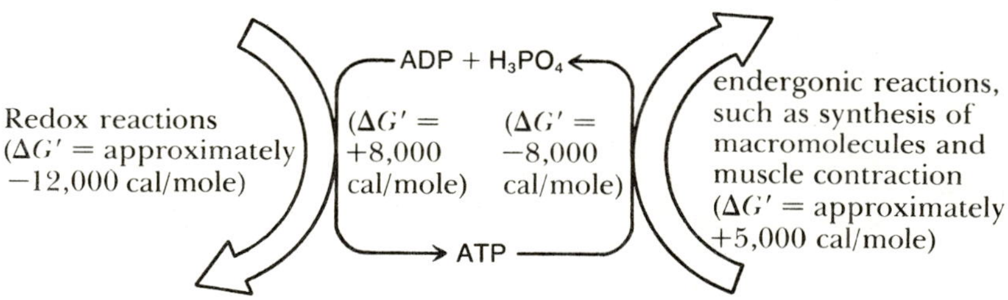

Some synthesis of ATP is coupled to reactions in glycolysis and the Krebs cycle. However, the amount of ATP produced is only approximately 12 percent (or one-ninth) as much as the amount produced by coupling of ATP synthesis to the electron transport chain.

4.62

A detailed look at the electron transport chain Because the electron transport chain is so important to those cells in which it occurs, it will be analyzed here in some detail. The following discussion will also be useful as a review of some of the concepts and terminology that you learned earlier in this unit on oxidation and reduction.

In the first step of the electron transport chain, electrons and protons are transferred from NADH to FAD. Write the two half reactions that symbolize this transfer.

4.61
no answer

4.63

Rewrite the correct half reactions requested in frame 4.62 using the "curved arrows" notation that was demonstrated in frame 4.35.

4.62

$NADH + H^+ \longrightarrow NAD^+ + 2\ e^- + 2\ H^+$

$FAD + 2\ H^+ + 2\ e^- \longrightarrow FADH_2$

4.64

In the second step of the electron transport chain, $FADH_2$ transfers two electrons to two molecules of cytochrome *b*. The two protons are released and go into the surrounding medium.

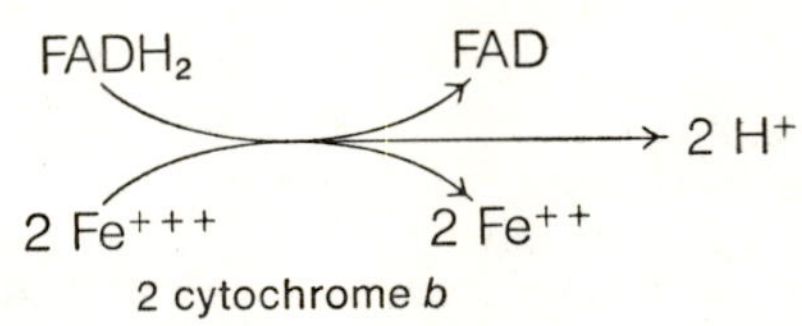

4.63

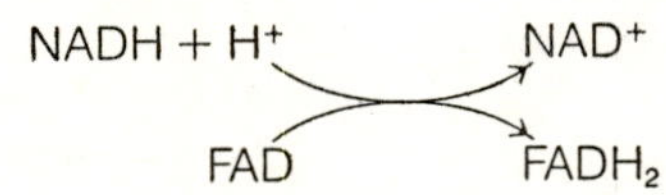

Note that in this reaction FAD is regenerated from $FADH_2$. FAD could now accept hydrogens from another $NADH + H^+$.

The first two steps of the electron transport chain can be diagramed as follows:

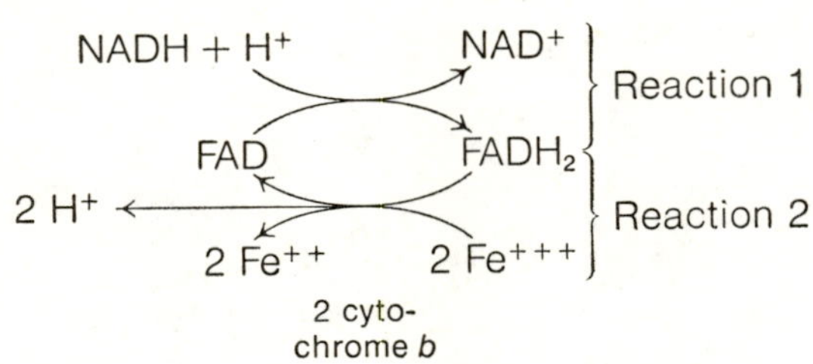

a. In reaction 1, FAD was
   ____oxidized.
   ____reduced.

b. In reaction 2, $FADH_2$ was
   ____oxidized.
   ____reduced.

4.65

After reaching cytochrome *b*, the two electrons are transferred to cytochrome *c*, then to cytochrome *a*, and finally to oxygen. In the last step, two protons from the surrounding medium join the reaction, and the final product is water:

4.64

a. reduced
b. oxidized

$$2\,e^- + \tfrac{1}{2}\,O_2 + 2\,H^+ \longrightarrow H_2O$$

Following is a partially completed diagram of the electron transport chain.[||] Fill in the blanks on this diagram.

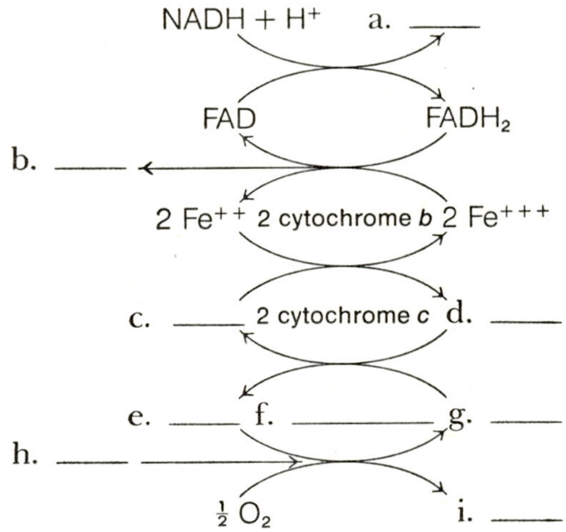

[||]Actually the electron transport chain is somewhat more complex than this diagram indicates, but none of the major steps has been omitted.

4.66 ______________________________

In the preceding diagram, trace the path of the electrons with a broken line.

4.67 ______________________________

The following statement is
____always true.
____sometimes true.
____never true.
When an oxidation occurs, a simultaneous reduction must also occur.

4.65
a. NAD^+
b. 2 H^+
c. 2 Fe^{+++}
d. 2 Fe^{++}
e. 2 Fe^{++}
f. 2 cytochrome *a*
g. 2 Fe^{+++}
h. 2 H^+
i. H_2O

4.66

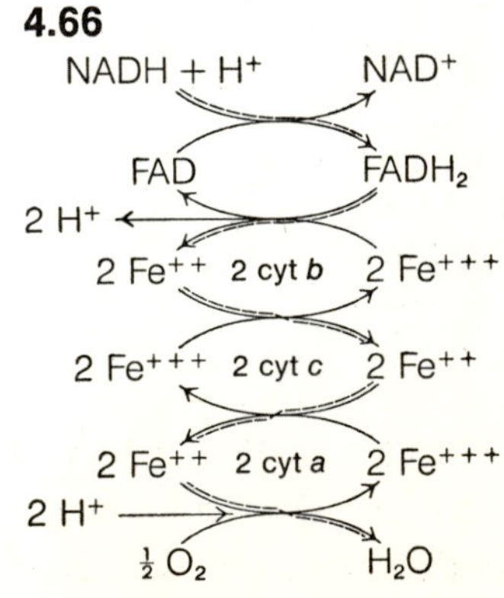

4.68 ______________________________

These two half reactions constitute the fourth step of the electron transport chain. Half reaction ____ is an oxidation and half reaction ____ is a reduction.

4.67
always true

1. 2 cytochrome *c*-Fe^{++} ⟶ 2 cytochrome *c*-Fe^{+++} + 2 e^-
2. 2 cytochrome *a*-Fe^{+++} + 2 e^- ⟶ 2 cytochrome *a*-Fe^{++}

4.69 ______________________________

Here is a list of the compounds that participate in the electron transport chain, arranged in order of increasing electron affinity:

4.68
Half reaction 1 is an oxidation; half reaction 2 is a reduction.

NAD^+	low electron affinity
FAD	↑ (*a*) ↓ (*b*)
cytochrome *b*	
cytochrome *c*	
cytochrome *a*	
O_2	high electron affinity

In which direction will electron transfers occur spontaneously?

____the direction indicated by arrow *a*
____the direction indicated by arrow *b*

4.70 ______________________________

a. E_0' stands for ______________________________.

b. Electron transfer from cytochrome *a* to oxygen is spontaneous in the cell. Therefore, which of these two compounds has the more positive value of E_0'?
____cytochrome *a*
____oxygen

4.69
the direction indicated by arrow *b*

4.71 ______________________________

When we determine the value of E_0' for a substance, to what substance do we compare its electron pressure or electron affinity?

4.70
a. standard reduction potential at *p*H = 7
b. oxygen

4.72 ______________________________

Which of the following two reactions would have a larger absolute value of $\Delta G'$?

4.71
hydrogen

____a reaction in which the E_0' of the oxidizing agent is similar in value to the E_0' of the reducing agent

____a reaction in which the E_0' of the oxidizing agent and the E_0' of the reducing agent differ greatly in value

4.73 ____

The changes in free energy that occur during electron transport are illustrated as follows. The broken arrow indicates the path of electrons.

4.72
a reaction in which the E_0' of the oxidizing agent and the E_0' of the reducing agent differ greatly in value

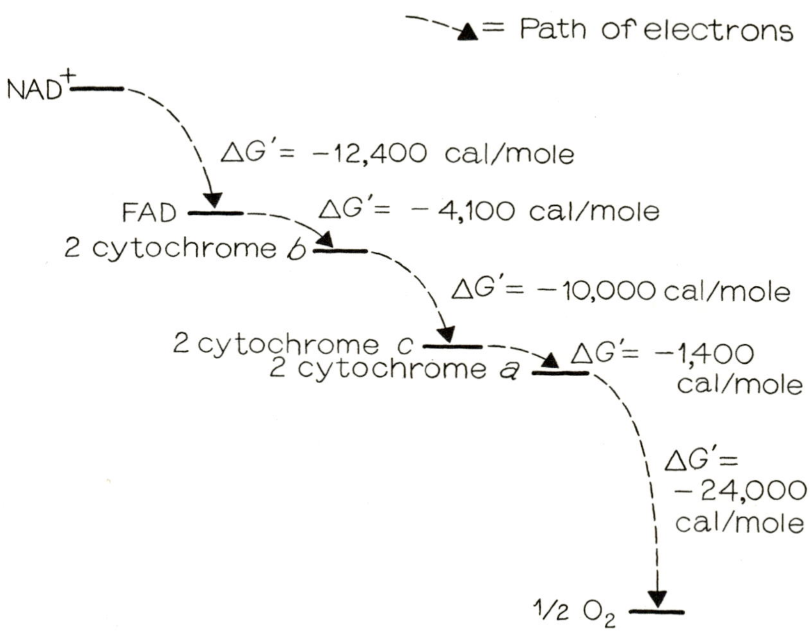

Which of the preceding reactions have values of $\Delta G'$ such that the reaction could be coupled to the synthesis of ATP, (shown as follows)?

$$ADP + H_3PO_4 \longrightarrow ATP + H_2O$$
$$(\Delta G' = +8{,}000 \text{ cal/mole})$$

4.74 ____

ATP production is coupled to three steps in electron transport, as shown in the following diagram:

4.73
electron transfers from:

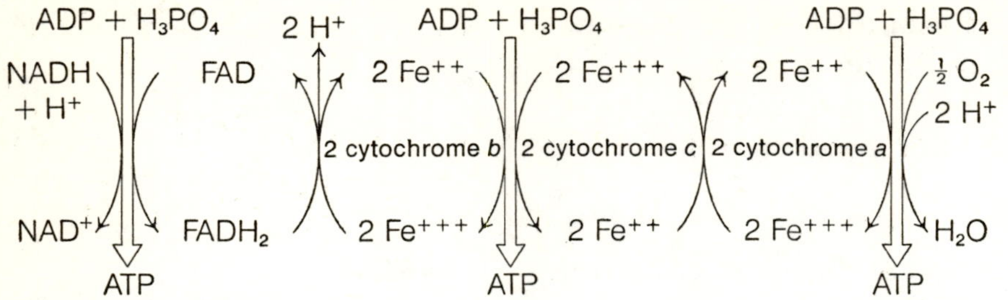

NAD^+ to FAD
cytochrome *b* to cytochrome *c*
cytochrome *a* to O_2

This production of ATP during electron transport is called oxidative phosphorylation.

Of what use to the cell is the ATP synthesized during oxidative phosphorylation?

4.74
The ATP is used to drive forward endergonic reactions vital to the cell.

seven
some alternatives to respiration

4.75

As you saw in Section Six, oxygen is the final electron acceptor in the electron transport chain. Thus electron transport cannot occur in cells that lack oxygen (for example, bacteria that live in an oxygen-free environment and cells of a muscle that has depleted its supply of oxygen during strenuous exercise). Although glycolysis does not produce as much ATP as respiration does, cells may use glycolysis as their source of ATP.

Answer the following questions by looking at the reactions of glycolysis shown on page 143.

a. Which reactions of glycolysis convert ADP to ATP?

b. For every molecule of glucose that goes through glycolysis, how many molecules of ATP are synthesized by these reactions?

c. Which reactions convert ATP to ADP?

d. For every molecule of glucose that goes through glycolysis, how many molecules of ATP are broken down to ADP by these reactions?

e. What is the net amount of ATP (in molecules) synthesized as one molecule of glucose goes through the glycolysis pathway?

4.76

In cells that carry out glycolysis but not the Krebs cycle or electron transport chain, one or more additional steps occur at the end of glycolysis. These reactions vary with the organism but serve the same purpose, to oxidize NADH to NAD^+.

To see why these steps are necessary, let us consider what would happen in a cell that carries out only glycolysis. During one step of the glycolytic pathway (reaction 6, shown here), NAD^+ is reduced to $NADH + H^+$:

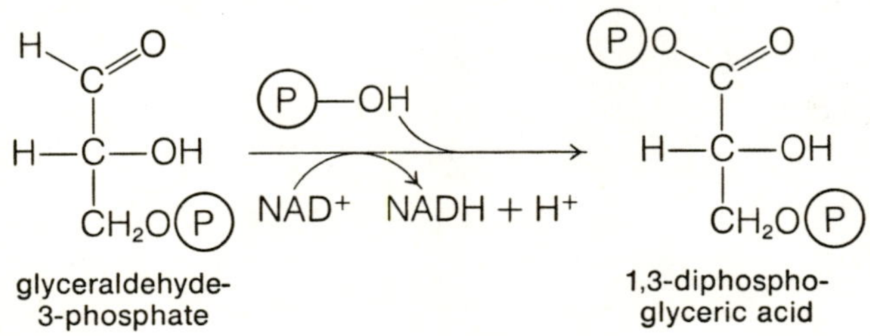

Any cell has only a finite amount of NAD^+ and NADH; therefore, all the NAD^+ in the cell may be reduced to NADH during a relatively brief stint of glycolysis. When this occurs, glycolysis comes to a halt for lack of NAD^+. Glycolysis could continue, however, if some additional reaction or reactions were to convert NADH back to NAD^+.

In animal cells and certain bacteria, the following reaction occurs at the end of glycolysis when oxygen is absent:

4.75

a. reactions 7 and 10
b. 4 (because 2 molecules of 3-carbon sugar are produced from each molecule of glucose)
c. reactions 1 and 3
d. 2
e. 2

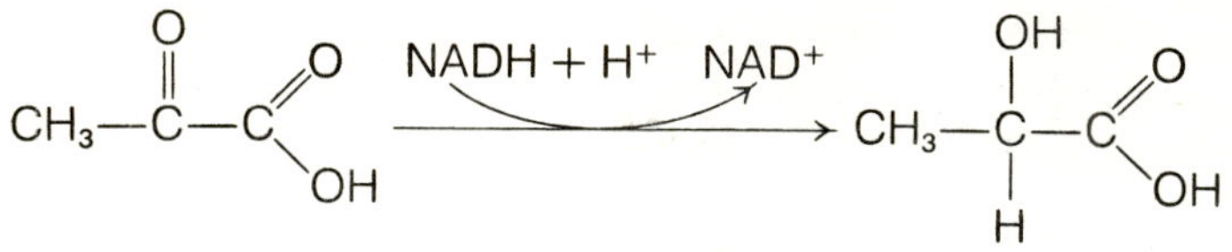

pyruvic acid | lactic acid

When oxygen is absent, the cells of higher plants, yeasts, and some bacteria employ the following two-step process at the end of glycolysis:

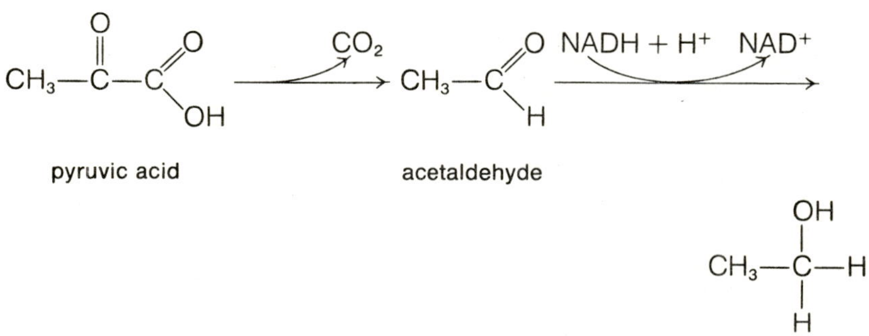

pyruvic acid | acetaldehyde | ethanol

---

review question

4.77

In cells that have a supply of oxygen and can carry out respiration, NAD^+ is reduced to $NADH + H^+$ during glycolysis. What prevents reduction of all the NAD^+ in the cell and consequent shutdown of glycolysis for lack of NAD^+?

4.76
no answer

4.78

Glycolysis, the two processes shown in frame 4.76, and respiration are the most widespread mechanisms by which cells break down food molecules in order to synthesize ATP from ADP and phosphate. However, they are by no means the only mechanisms. Many different pathways have evolved, particularly among the micro-organisms, and living things are able to use a wide variety of substances as food. Nevertheless, the reason for cellular breakdown of food molecules is universal.

4.77
NADH is oxidized back to NAD^+ in the first step of the electron transport chain.

Exergonic reactions in the breakdown process are coupled to the synthesis of ATP from ADP and phosphate. In turn, the conversion of ATP to ADP and phosphate is coupled to vital cell reactions that are endergonic and would not otherwise occur.

You have now completed Unit Four.

help frames for unit four

4.79
4.23

The atomic structure of the hydrogen atom is as follows:

4.78
no answer

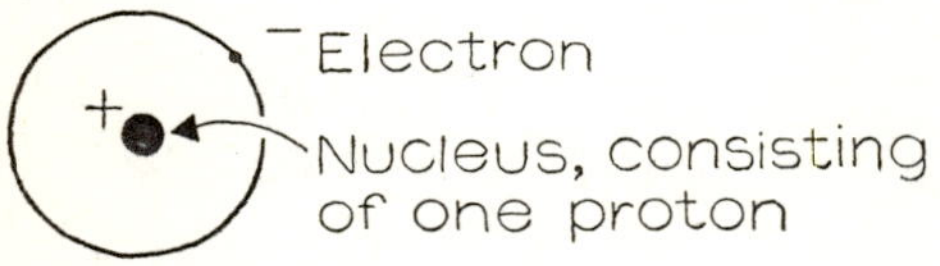

An H^+ ion is simply a hydrogen atom that has lost one electron. If one electron is removed from the hydrogen atom, all that remains is a proton. Thus H^+ ions and protons are the same thing.

Return to frame 4.23.

4.80
4.36

The path taken by the electrons in the following reaction is represented by a dotted line.

4.79
no answer

Return to frame 4.36.

4.81
4.51

In both reactions in frame 4.51, the electron acceptor is oxygen, so the difference must be in the electron donor. Which

4.80
no answer

reaction contains the substance with the greater tendency to donate electrons?

Return to frame 4.51.

4.82

4.29

The structure of FAD (flavin adenine dinucleotide) A molecule of FAD has the following structure:

4.81
reaction 2

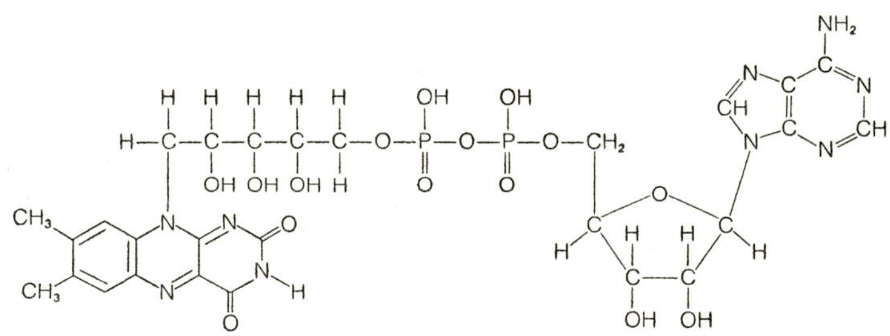

The separate components of FAD are identified in the following diagram:

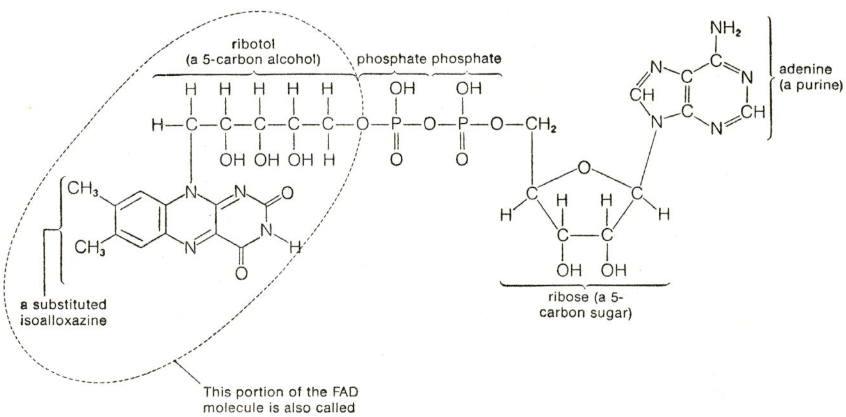

$FADH_2$ differs from FAD only in the substituted isoalloxazine, which in $FADH_2$ has the structure:

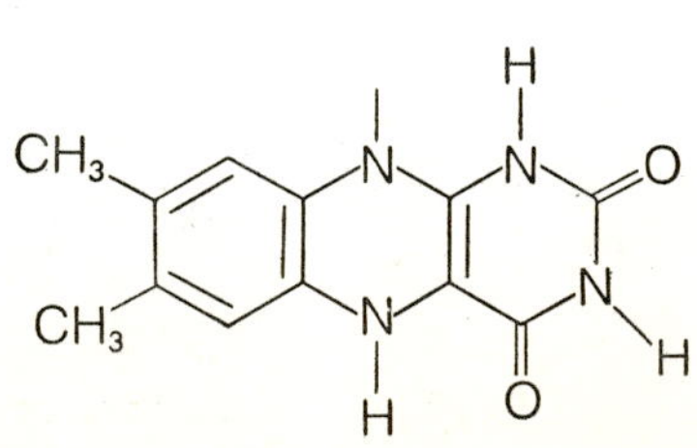

● Go to frame 4.83.

4.83

The structure of NAD$^+$ (nicotinamide adenine dinucleotide) NAD$^+$ consists of the following components: **4.82** no answer

nicotinamide
|
ribose
|
phosphate
|
phosphate
|
ribose
|
adenine

The structure of NAD$^+$ is:

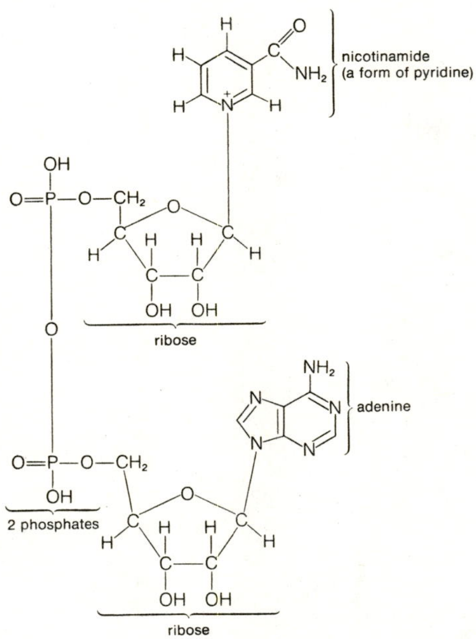

When NAD$^+$ is reduced to NADH + H$^+$, the nicotinamide portion of the molecule has this structure:

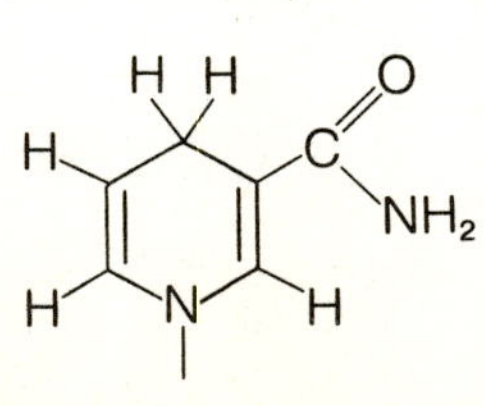

● Return to frame 4.29. **4.83** no answer

reactions in glycolysis

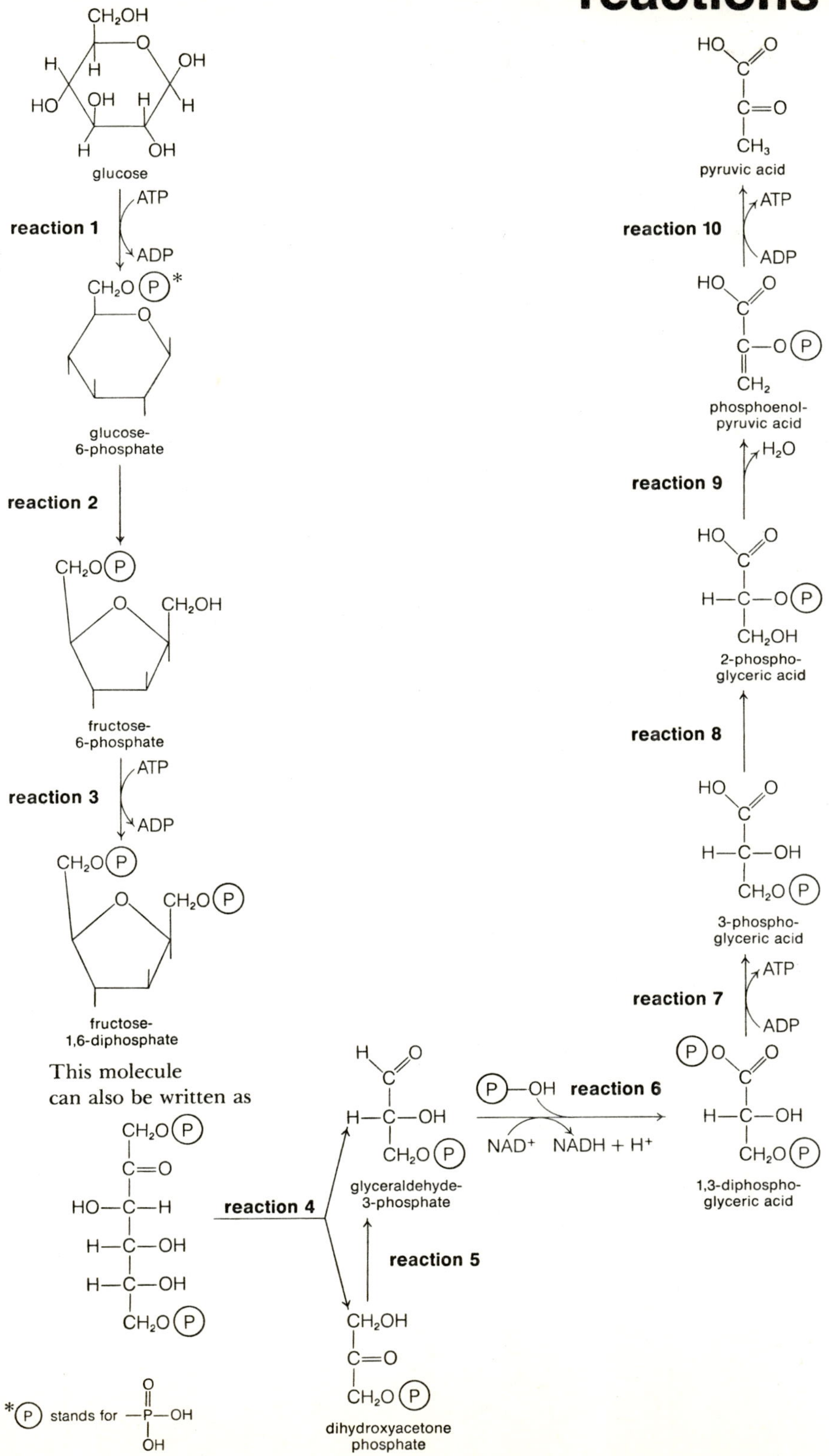

reactions in krebs cycle

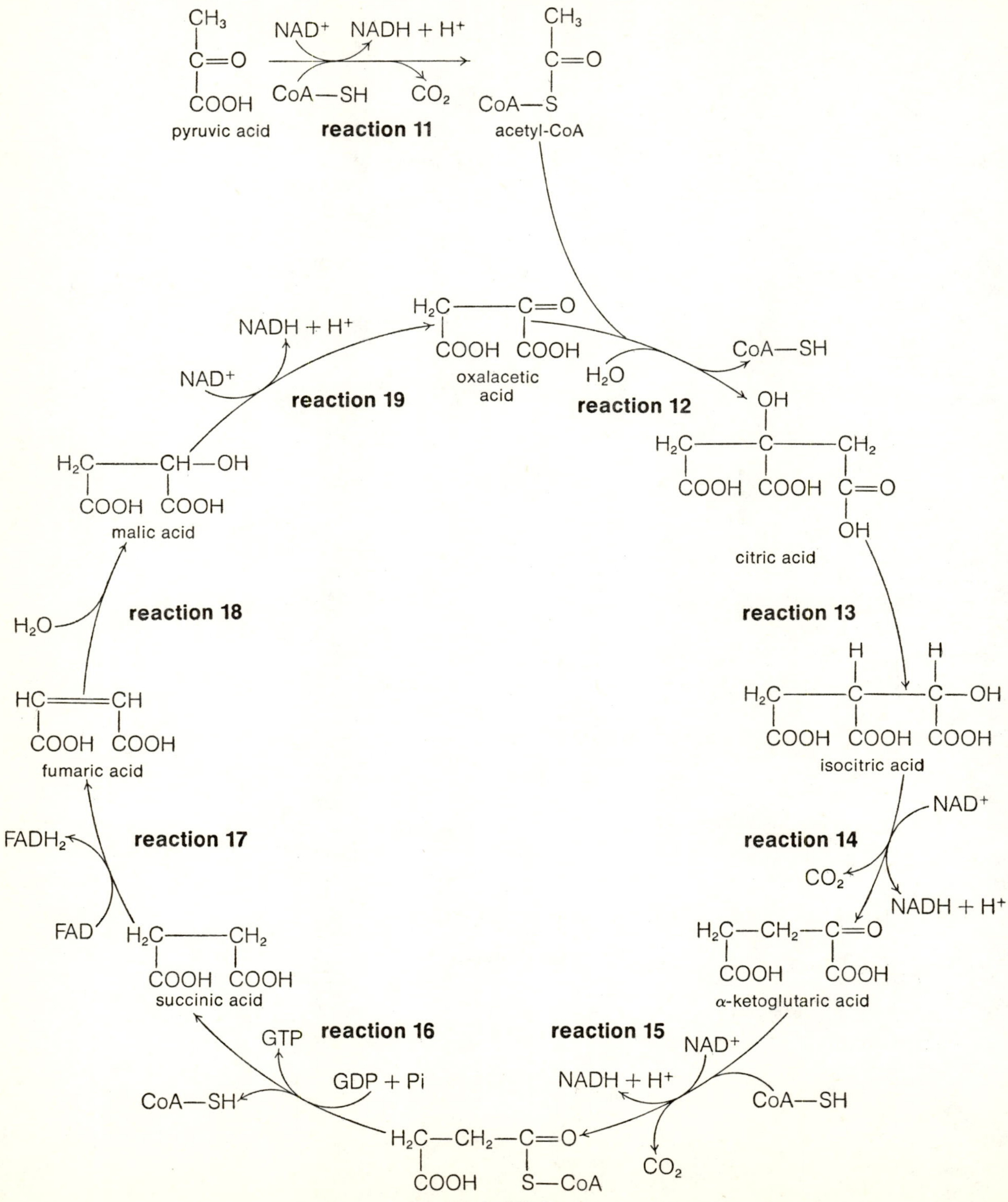

glossary

absolute value the absolute value of a positive number equals that number. The absolute value of a negative number equals −1 times that number.

active site of an enzyme the cleft on the surface of the enzyme that contains catalytic groups and to which the substrate binds.

ADP *See* ATP.

amino acid a compound having the following formula and structure:

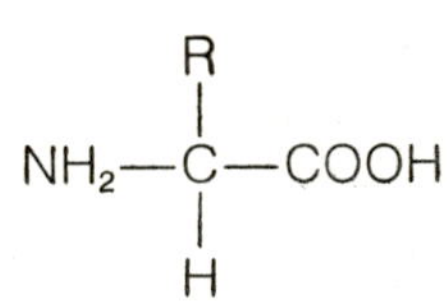

Amino acids are the subunits of which proteins are composed. Cells contain approximately twenty different amino acids.

ATP adenosine triphosphate. Its structure is as follows:

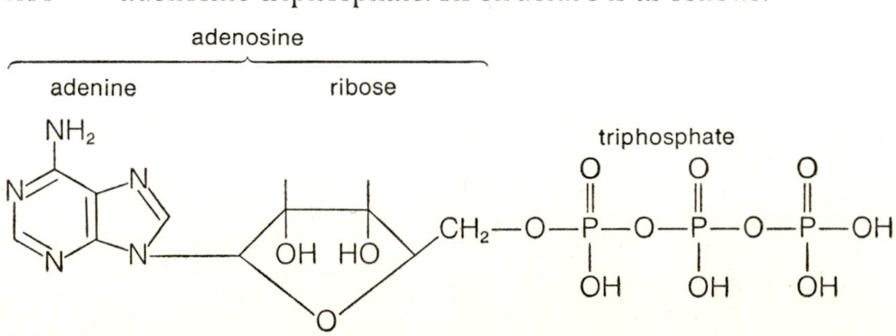

Adenosine is formed by joining ribose, a sugar, to adenine, a nitrogenous base. ADP, or adenosine diphosphate, has the same structure as ATP except that it contains only two phosphate groups.

balanced redox reaction a reaction in which the total charge on the reactants equals the total charge on the products.

carbohydrate a compound having the general formula $(CH_2O)_n$. Examples of carbohydrates are sugars, starches, and cellulose.

catalyst a substance that increases the rate of a chemical reaction without being consumed or altered permanently by the reaction.

common intermediate a substance that is a product of one reaction and a reactant of another reaction.

coupled reactions a series of two or more reactions in which the endergonic reaction(s) is(are) driven forward by the exergonic reaction(s).

derivative a substance derived from some other substance through chemical change. A derivative of glucose is a compound that was formed as a result of one or more changes in the glucose molecule; for example, glucose-6-phosphate, phosphoglyceric acid, pyruvic acid, citric acid, and all other compounds in glycolysis and the Krebs cycle containing atoms that originally were part of a glucose molecule.

disordered situation a situation that can be obtained in relatively many ways, or a situation in which different regions of the same system are not easily distinguished.

E_0 reduction potential under standard conditions.

E_0' reduction potential at *p*H 7, other conditions being standard.

electron transport chain a series of reactions in which electrons are passed from one electron acceptor to another after being removed from glucose and its derivatives. Synthesis of ATP is coupled to some of the reactions in the electron transport chain.

endergonic having a positive ΔG; nonspontaneous.

enzyme a protein molecule that functions in the cell as a catalyst.

equilibrium in a chemical system, the state in which no further change occurs in the concentrations of reactants and products (because the forward reaction and the reverse reaction are occurring at the same rate).

ester a compound formed as a result of a reaction between an alcohol (R′—OH) and a carboxylic acid,

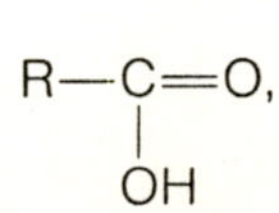

or phosphoric acid. Carboxylic esters have the general formula

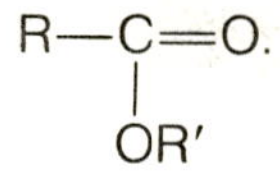

free energy a measurable property of a nonisolated system. Free energy of a system decreases as the total disorder of (nonisolated system plus its surroundings) increases. In other words, changes that are spontaneous are accompanied by a decrease in free energy.

Functional group a small group of atoms that is part of an organic compound and that differs in composition from simple hydrocarbons. Examples are —OH, —NH_2, —C=C—, —C=O.

ΔG change in free energy; free energy of products minus free energy of reactants.

$\Delta G°$ change in free energy under standard conditions.

$\Delta G'$ change in free energy at *p*H 7, other conditions being standard.

glycolysis the series of reactions by which carbohydrates such as glucose are converted to pyruvic acid. See diagram on p. 143.

half reaction an equation that shows an oxidation without showing what substance accepts the electrons or that shows a reduction without showing what substance donates the electrons. For example:

$$\text{Fe (metal)} \longrightarrow \text{Fe}^{++} + 2\,e^-$$

hydrolysis the splitting of a molecule into two or more parts by the addition of water. For example:

sucrose + H_2O ⟶ glucose + fructose

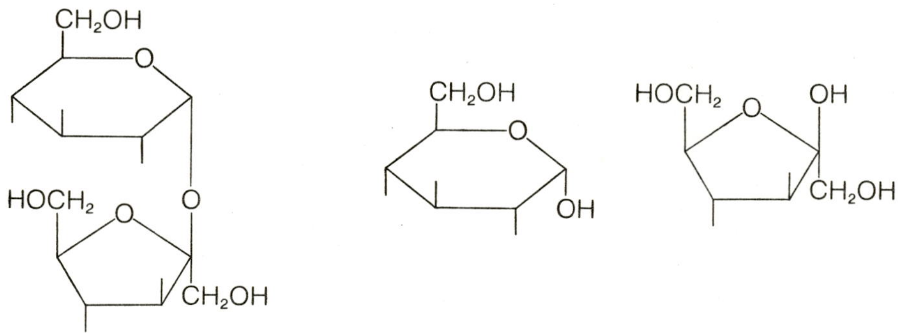

ion an atom or group of atoms that is electrically charged because of an excess or deficiency of electrons; for example, Cl^-, Na^+, NH_4^+.

isolated system *See* System.

kinetic energy energy involved in motion.

Krebs cycle (also known as citric acid cycle and tricarboxylic acid cycle) the part of the respiration pathway in which pyruvic acid is broken down into carbon dioxide. See diagram on p. 144.

lipids fatty substances; a class of complex organic compounds that are insoluble in water but soluble in such organic solvents as acetone.

macromolecule generally a polymer of a few or many subunits. Proteins, polysaccharides, and nucleic acids, all of which are important constituents of the cell, are examples of macromolecules. The repeating subunits of these three macromolecules are amino acids, simple sugars, and nucleotides, respectively.

membrane a thin sheet (thought to be approximately four molecules thick), composed of lipid and protein molecules in an orderly array. Such a membrane surrounds all cells, and a double membrane surrounds the nuclei of eucaryotic cells. Membranes are also important components of many cell organelles, such as chloroplasts and mitochondria.

mitochondria the organelles in which the reactions of respiration, with the exception of glycolysis, occur. An average cell contains approximately 500 mitochondria. (The cells of bacteria and blue-green algae do not contain mitochondria. Reactions in these cells are thought to take place on infoldings of the membrane that surrounds each cell.) A mitochondrion can be diagramed as follows:

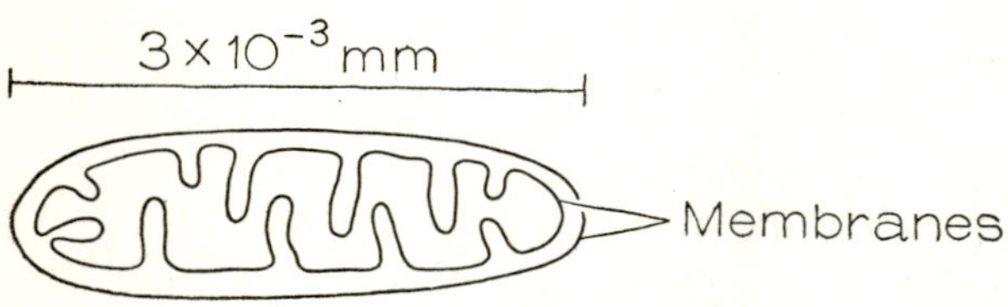

monomer *See* Polymer.

nucleic acid a linear polymer of nucleotides; a macromolecule consisting of a sugar-phosphate backbone to which are attached nitrogenous bases such as adenine. (See the definition of ATP for the structure of adenine.) A nucleic acid can be diagramed as follows:

—sugar—PO_4—sugar—PO_4—sugar—PO_4—sugar—PO_4—
| | | |
Base Base Base Base

DNA, the carrier of genetic information, is a nucleic acid.

orderly situation a situation that can be obtained only in relatively few ways, or a situation in which different regions of the same system are easily distinguished.

organelle an organized structure within the cell, such as nucleus, chloroplast, mitochondrion.

oxidation loss of electrons or hydrogen atoms.

oxidation-reduction reaction a reaction in which electrons are transferred from one substance to another; a redox reaction.

oxidizing agent the substance that accepts electrons from another substance, thus oxidizing it.

pathway a sequence of reactions by which one compound is converted to another.

PEP phosphoenolpyruvic acid:

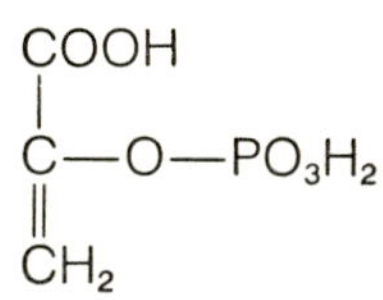

pH a way of expressing the concentration of hydrogen ions in a solution. *p*H is defined as $-1 \times \log_{10} [H^+]$. A *p*H of 7 is neutral.

polymer a molecule composed of repeating subunits, called monomers.

polypeptide a linear molecule composed of amino acids joined by covalent bonds.

polysaccharide a molecule composed of sugar molecules that are linked together. Starch and cellulose are examples of polysaccharides.

product *See* Reactant.

protein a molecule composed of one or more polypeptides folded in a complex manner and held together by weak chemical bonds.

proton the H^+ ion. See frame 4.79.

reactant a substance that is present at the beginning of a chemical reaction and that undergoes a chemical change during the reaction. In a chemical equation, reactants are usually shown on the left side of the arrow:

$$\underbrace{2\ H_2 + O_2}_{\text{Reactants}} \longrightarrow \underbrace{2\ H_2O}_{\text{Products}}$$

redox reaction a reaction in which electrons are transferred from one substance to another; an oxidation-reduction reaction.

reducing agent a substance that donates electrons to another substance, thus reducing it.

reduction gain of electrons or hydrogen atoms.

reduction potential E_0; a measure of the ability of a substance to accept electrons.

respiration the series of chemical reactions in the cell whereby foodstuff molecules react with oxygen, producing carbon dioxide and water; the reactions of the Krebs cycle and the electron transport chain.

spontaneous reaction a reaction that occurs without such outside intervention as the addition of matter or energy. The word "spontaneous" does not imply that a reaction happens at a rapid or even at a noticeable rate.

standard conditions the following conditions are defined as standard: a temperature of 25° C, gases at one atmosphere pressure, dissolved substances at a concentration of one mole/liter (consequently, pH = 0), and pure liquids and solids in their most stable form at one atmosphere and 25° C.

substrate the reactant(s) of an enzyme-catalyzed reaction.

system the particular collection of matter on which we are focusing our attention. The rest of the universe comprises the surroundings of the system. An isolated system is one that does not interact with any other systems. (For example, an isolated system does not exchange energy or matter with any other systems.)

final test

The questions in this test cover the main points of the program. The title or numbers in parentheses above each question refer to the section or frame(s) where the topic covered in that question is discussed. You will find the answers to the questions on p. 162.

(1.15)

1. Will the following change occur spontaneously? (Assume that this is an isolated system and that kinetic and potential energy are not altered during the change.)
   ____yes
   ____no

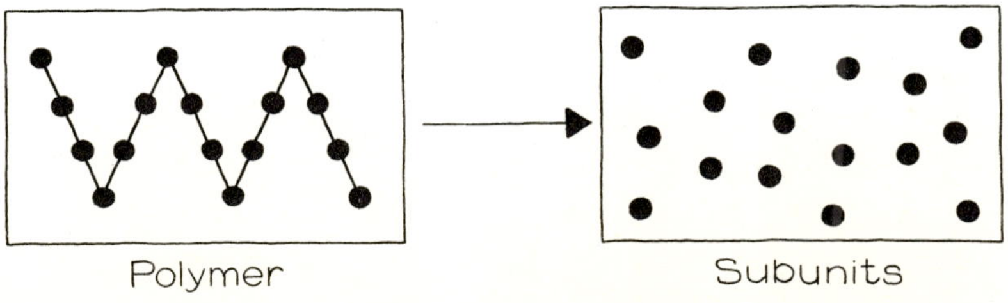

(1.28)

2. Will the following change occur spontaneously?

_____yes

_____no

Before change: disorder of mixture in the flask (a non-isolated system) = 6; disorder of surroundings = 9.

After change: disorder of mixture in the flask = 10; disorder of surroundings = 4.

(1.36)

3. Choose the correct phrases to complete the following statement: The free energy of a nonisolated system will

_____increase

_____decrease

when there is an increase in the disorder of the

_____nonisolated system.

_____surroundings.

_____nonisolated system + the surroundings.

(1.33)

4. As an organism grows, it increases in size and complexity and thus decreases in disorder. Which statement explains how respiration makes possible a decrease in the disorder of the organism?

_____

a. During respiration, simple molecules are joined to form polymers of great complexity, thus decreasing the disorder of the organism.

b. Respiration results in the breakdown of food molecules into simpler structures that are the building blocks for complex compounds, such as proteins. Synthesis of complex compounds constitutes a decrease in disorder.

c. During respiration, complex food molecules are changed to simpler CO_2 and H_2O molecules, which are released to the surroundings. This causes an increase in the disorder of the surroundings that is greater than the decrease in the disorder of the organism caused by growth.
d. None of these statements offers a suitable explanation.

(1.40, 1.55, 1.56)

5. Decide to which reaction on the right the symbol on the left applies and write the letter of that reaction in the space provided.

$\Delta G°$_____
ΔG _____
$\Delta G'$_____

a. Reaction taking place under cellular conditions.
b. Reaction taking place under unspecified or unknown conditions.
c. Reaction taking place at 0° C and pH = 0.
d. Reaction taking place at 25° C, pH = 0, and concentrations = one molar.

(1.59)

6. Draw a free energy diagram for the following reaction:

$$\text{glucose} + \text{fructose} \longrightarrow \text{sucrose} + H_2O \quad (\Delta G = +5{,}500 \text{ cal/mole})$$

(1.41)

7. Will the reaction in question 6 occur spontaneously?

_____yes
_____no

(1.43)

8. What is the $\Delta G'$ for the reverse of the following reaction?

_____It is not possible to say from the available data.
_____The value is _____________ cal/mole.

$$X + Y \longrightarrow Z + W$$

$$(\Delta G' = +320 \text{ cal/mole})$$

(1.44)

9. Indicate whether the ΔG's of the following pairs of reactions will be the same or different. Temperatures appear at right.

a. glucose + fructose ⟶ sucrose + H_2O (20° C)
(.25 molar) (.25 molar) (.75 molar)

glucose + fructose ⟶ sucrose + H_2O (20° C)
(.25 molar) (.75 molar) (.25 molar)

_____same
_____different

b. glucose ⟶ 2 lactic acid (35° C)
(1 molar) (1 molar)

glucose ⟶ (4H released) 2 pyruvic acid ⟶ (4H added) 2 lactic acid (35° C)
(1 molar) (1 molar) (1 molar)

_____same
_____different

c. fumaric acid ⟶ malic acid + H_2O (20° C)
(10^{-2} molar) (10^{-4} molar)

fumaric acid ⟶ malic acid + H_2O (25° C)
(10^{-2} molar) (10^{-4} molar)

_____same
_____different

d. fumaric acid —(in the presence of fumarase, an enzyme that accelerates the reaction)⟶ malic acid + H_2O (20° C)
(1 molar) (.5 molar)

fumaric acid —(no fumarase present)⟶ malic acid + H_2O (20° C)
(1 molar) (.5 molar)

_____same
_____different

(1.57)
10. If a reaction is endergonic, it will
_____occur spontaneously.
_____not occur spontaneously.

(2.24)
11. Following are four sets of reactions. In which set(s) can the reactions be coupled to each other? ($\Delta G'$ in calories/mole appears at right.)

____

1. ATP + H_2O ⟶ ADP + H_3PO_4 −8,000
fructose + H_3PO_4 ⟶ fructose-6-phosphate + H_2O +3,800

2. pyruvic acid ⟶ acetaldehyde + CO_2 −14,400
pyruvic acid + H_3PO_4 ⟶ phosphoenolpyruvic acid + H_2O +12,800

3. glucose-1-phosphate + H_2O ⟶ glucose + H_3PO_4 −5,000
ADP + H_3PO_4 ⟶ ATP + H_2O +8,000

4. acetyl phosphate + H_2O ⟶ acetic acid + H_3PO_4 −10,000
glucose + H_3PO_4 ⟶ glucose-6-phosphate + H_2O +3,300
glucose-6-phosphate ⟶ fructose-6-phosphate +400

(2.25)
12. Rewrite the following pair of coupled reactions in the conventional notation used for coupled reactions:

UDP-glucose ⟶ UDP + glucose
$\text{glycogen}_{(n)}$ + glucose ⟶ $\text{glycogen}_{(n+1)}$

(2.25, 2.28)

13. Which of the following represent(s) coupled reactions?

____

1. lactose + H_2O ⟶ glucose + galactose

2. pyruvic acid $\xrightarrow{CO_2}$ acetaldehyde $\xrightarrow{2\ H}$ ethanol

3. acetyl phosphate + glycerol ⟶ acetic acid + glycerol phosphate

4. 3-phosphoglyceric acid ⇌ 2-phosphoglyceric acid

(2.05)

14. Following the pair of coupled reactions that appear below is a list of possible reasons to explain why reaction 2 proceeds. Which explanation is most correct?

____

1. fructose-6-phosphate + H_2O ⟶ fructose + H_3PO_4

$(\Delta G' = -3{,}800 \text{ cal/mole})$

2. glycerol + H_3PO_4 ⟶ glycerol-1-phosphate + H_2O

$(\Delta G' = +2{,}300 \text{ cal/mole})$

a. Reaction 2 is forced to proceed because reaction 1 consumes a product of reaction 2.
b. Reaction 2 proceeds because it uses up glycerol, which is in abundant supply in the reaction mixture.
c. Reaction 2 proceeds because reaction 1 spontaneously floods the system with a reactant of reaction 2.
d. Reaction 2 proceeds because its $\Delta G'$ is positive.

(3.03)

15. The following reaction occurs

_____at a relatively fast rate.

_____at a relatively slow rate.

_____not enough information is given.

$$\text{fumaric acid} + H_2O \longrightarrow \text{malic acid}$$
$$(\Delta G' = -750 \text{ cal/mole})$$

("An Overview of Cellular Metabolism")
16. "A spontaneous reaction proceeds at a rapid rate." This statement is

_____always true.
_____sometimes true.
_____never true.

(3.05)
17. Describe what must take place between two reactant molecules before they can become products.

(3.12)
18. Which line in the following graph represents the free energy of activation?

_____

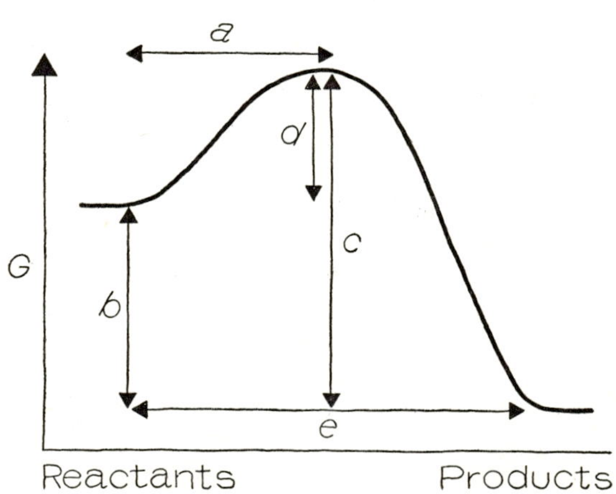

(3.20)
19. Which of the following statements explains why a reaction proceeds faster if the temperature is raised? (Choose one or more.)

_____

a. At high temperatures, there are more molecules that possess so much kinetic energy that converting some of it to potential energy does not decrease disorder.
b. At high temperatures, the molecules are moving faster and the system is more disordered.
c. More collisions occur per second at high temperatures than at low temperatures.
d. The difference in free energy between reactants and products is less at high temperatures than at low temperatures.

(3.22)
20. By which method(s) can catalysts accelerate reactions?

____

a. Increase the stability of reactants.
b. Increase the stability of products.
c. Increase the stability of the transition state.
d. Decrease the stability of reactants.
e. Decrease the stability of products.
f. Decrease the stability of the transition state.

(3.21, 3.22)
21. Which statement(s) regarding catalysts is(are) true?

____

a. Catalysts lower the ΔG of a reaction.
b. Catalysts may be altered permanently or used up by the reactions they catalyze.
c. Catalysts increase the kinetic energy of the reactant molecules.
d. Catalysts lower the free energy of activation of a reaction.

(3.29)
22. Which of the following statements is(are) true?

____

a. An enzyme is the opposite of a catalyst.
b. A catalyst is a kind of enzyme.
c. An enzyme is a kind of catalyst.
d. Enzyme and catalyst are mutually exclusive categories.
e. Some, but not all, enzymes are catalysts.

(3.29–3.31)
23. Which statement(s) about enzymes is(are) true?

____

a. One enzyme usually catalyzes many reactions.
b. Enzymes are composed mostly of lipid.
c. Enzymes are composed mostly of carbohydrate.
d. The three-dimensional structure of an enzyme is a thread coiled into a helix.
e. The three-dimensional structure of an enzyme is a thread folded many times into a globular shape.
f. Chemical bonds are not formed between an enzyme and its substrate.

(3.31–3.34)
24. By which method(s) can enzymes accelerate reactions?

____

a. As the substrate binds to the enzyme, the shape of the substrate is distorted and bonds in the substrate are thus weakened.
b. When an enzyme is present, the reaction can proceed by a pathway that contains more steps than the nonenzymatic pathway. In each step the free energy of activation is lower than the maximum free energy of activation in the nonenzymatic pathway.
c. When an enzyme is present, the reaction can proceed by a pathway that contains more steps than the nonenzymatic pathway. In each step the free energy of activation is higher than the maximum free energy of activation in the nonenzymatic pathway.
d. An enzyme contains catalytic groups that are favorably arranged in space and that come into contact with and interact with the substrate.
e. None of these.

(4.02, 4.22)
25. Which of the following four reactions are redox reactions? (Choose one or more.)

____

a. dehydroascorbic acid + 2 H atoms $\longrightarrow$ ascorbic acid

b. $2\,Fe + O_2 \longrightarrow 2\,Fe^{++}O^{--}$

c. $H^+Cl^- + Ag^+NO_3^- \longrightarrow Ag^+Cl^- + H^+NO_3^-$

d. fumaric acid + $H_2O \longrightarrow$ malic acid

(4.02)

26. The following reaction is
_____an oxidation.
_____a reduction.

$$Fe^{++} \longrightarrow Fe^{+++} + e^-$$

(4.17, 4.29)

27. Indicate whether the form given for each of the following compounds is the oxidized or the reduced form of that compound.

a. FAD
_____oxidized
_____reduced

b. NAD^+
_____oxidized
_____reduced

c. cytochrome *a*-Fe^{++}
_____oxidized
_____reduced

(4.38)

28. In the following reaction, the oxidizing agent is
_____________________.

$$FADH_2 + 2 \text{ cytochrome } c\text{-}Fe^{+++} \longrightarrow FAD + 2 \text{ cytochrome } c\text{-}Fe^{++}$$

(4.29)

29. Complete the following equation:

$$NAD^+ + H_2O \longrightarrow \tfrac{1}{2}\, O_2 + \underline{\hspace{4cm}}$$

(4.34, 4.35)

30. Rewrite the following as two half reactions:

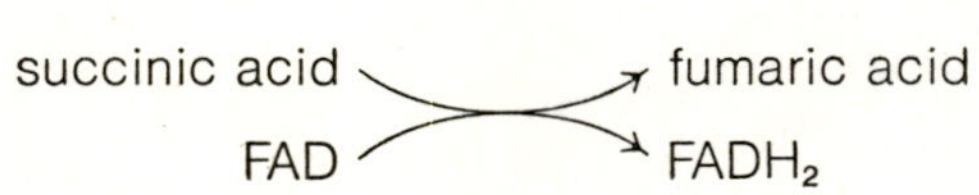

(4.37)
31. In the following reaction, electrons are transferred from
______cytochrome *a*-Fe^{+++}
______cytochrome *a*-Fe^{++}
______ethanol
______acetaldehyde + 2 H^{+}
and ultimately become part of
______cytochrome *a*-Fe^{+++}
______cytochrome *a*-Fe^{++}
______ethanol
______acetaldehyde + 2 H^{+}

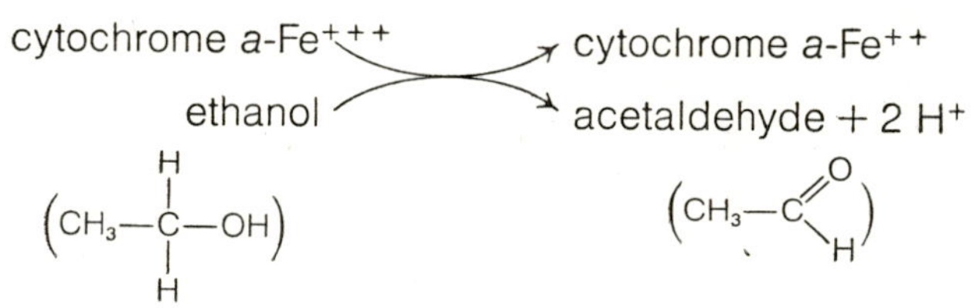

(4.49)
32. Electrons flow spontaneously
______from compounds with low electron affinity to compounds with high electron affinity.
______from compounds with high electron affinity to compounds with low electron affinity.

(4.49)
33. Which compound has the more positive E_0?
______a compound with high electron affinity
______a compound with low electron affinity

(4.58–4.65)
34. Which of the following three statement(s) is(are) true?

______

a. During glycolysis and the Krebs cycle, electrons are transferred from reduced forms of NAD^{+} and FAD to molecules derived from glucose (for example, pyruvic acid).
b. Compounds in the electron transport chain are arranged in order of decreasing electron affinity.
c. FAD, NAD^{+}, and the cytochromes participate in the electron transport chain.

answers to final test questions

1. yes
2. no
3. decrease; nonisolated system + the surroundings
4. c
5. $\Delta G°$: d
   ΔG : b
   $\Delta G'$: a
6. 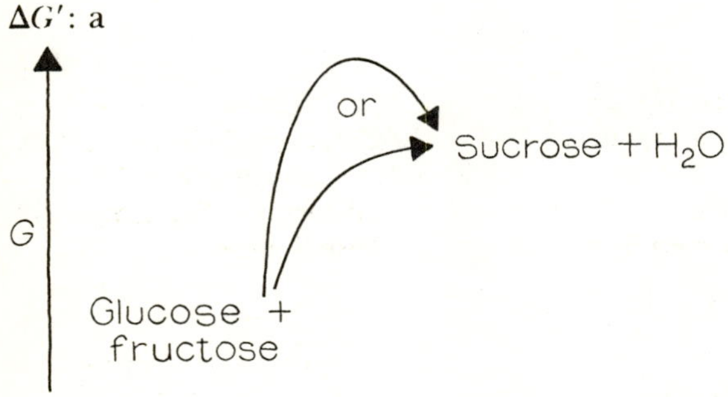

7. no
8. the value is −320 cal/mole
9. different; same; different; same
10. not occur spontaneously
11. 1, 4
12. 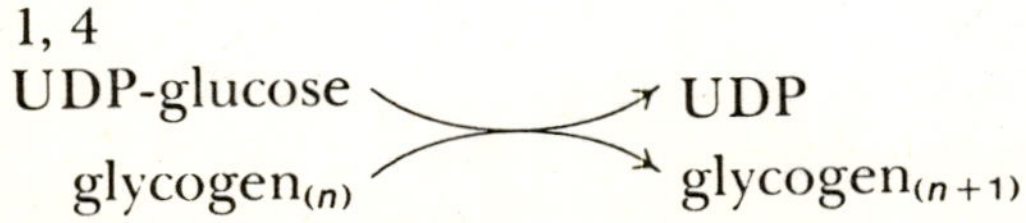

13. 2, 3
14. c
15. not enough information is given
16. sometimes true
17. Reactant molecules must collide, sometimes in a certain orientation. Old bonds must be disrupted or broken and new bonds must form; in other words, the transition state must be formed.
18. d
19. a,c
20. c,d
21. d
22. c
23. e
24. a,b,d
25. a,b
26. an oxidation
27. a. oxidized
    b. oxidized
    c. reduced
28. 2 cytochrome *c*-Fe^{+++}
29. NADH + H^+
30. succinic acid $\longrightarrow$ fumaric acid + 2 H^+ + 2 e^-
    FAD + 2 H^+ + 2 e^- $\longrightarrow$ $FADH_2$
31. ethanol, cytochrome *a*-Fe^{++}
32. from compounds with low electron affinity to compounds with high electron affinity
33. a compound with high electron affinity
34. c

index